HISTORY

BREAKING HISTORY

LOST AMERICA

LOST AMERICA

THE VIKING CITY OF NEW ENGLAND · ROANOKE · LOVE CANAL · ROUTE 66 · NEVERLAND

VANISHED CIVILIZATIONS, ABANDONED TOWNS, AND ROADSIDE ATTRACTIONS

DON RAUF

Guilford, Connecticut

An imprint of Globe Pequot
An imprint of The Rowman & Littlefield Publishing Group, Inc.
4501 Forbes Blvd., Ste. 200
Lanham, MD 20706
www.rowman.com

Distributed by NATIONAL BOOK NETWORK

British Library Cataloguing in Publication Information available

Library of Congress Cataloging-in-Publication Data available

ISBN 978-1-4930-3396-6 (hardcover)
ISBN 978-1-4930-3397-3 (e-book)

♾™ The paper used in this publication meets the minimum requirements of American
National Standard for Information Sciences—Permanence of Paper for Printed Library
Materials, ANSI/NISO Z39.48-1992.

Printed in the United States of America

To my mother, Charlotte, the school librarian who gave me a deep love of books and history.

❧

"A people without the knowledge of their past history, origin and culture is like a tree without roots."

—Marcus Garvey

CONTENTS

INTRODUCTION 9

PART I: LOST KINGDOMS OF EARLY AMERICA:
The Mythical and the Real 11
    CAHOKIA: North America's First Great City, **12**
    QUIVIRA and the Seven Golden Cities of Cibola, **33**
    THE FOUNTAIN OF YOUTH, **44**
    NORUMBEGA: The Viking City of New England, **47**

PART II: ABANDONED AMERICA:
Communities that Flourished and Faded 59
    THE VAST ANASAZI CIVILIZATION
      OF THE SOUTHWEST, **61**
    ROANOKE: America's First Colony, **72**
    BODIE: One of America's Most Famous Ghost Towns, **96**
    NORTH BROTHER ISLAND: New York Center for
      Quarantining the Sick, **113**

PART III: MOTHER NATURE STRIKES:
Towns Lost to Natural Disaster 129
    RUDDOCK, NAPTON, AND FRENIER, LOUISIANA:
      Wiped Out by Hurricane, **131**
    GONE FOR GOOD—OTHER US TOWNS DESTROYED BY
      HURRICANES: Isle Derniere, Indianola, and Hog
      Island, **142**
    VANPORT: Flood Washes Away Oregon's
      Second-Largest City, **147**

OTHER DROWNED US TOWNS: Old Cahawba, Alabama,
        The Catskills Underwater, **158**

PART IV: MAN-MADE DISASTERS:
Towns Too Polluted to Survive.......................... **163**
        LOVE CANAL: America's First Toxic Ghost Town, **165**
        CENTRALIA: The Town that Never Stopped Burning, **174**
        GILMAN: A Deadly Water Supply Drives a Town to
        Extinction, **194**

PART V: LOST ICONIC AMERICA:
Famous Sites that Have Disappeared **205**
        ROUTE 66: America's Road to Nowhere, **207**
        TWO OTHER LOST HIGHWAYS : The Oregon Trail,
        The Lincoln Highway, **220**
        OTHER LOST ICONS: Disappearing Drive-Ins,
        Adieu to Drive-In Diners, Too, Abandoned
        Amusement Parks, **222**

BIBLIOGRAPHY **228**
ABOUT THE AUTHOR........................... **236**
ACKNOWLEDGMENTS........................... **236**

INTRODUCTION

When it comes to America, the one thing that is constant is change. As the country moves relentlessly forward and continues to evolve, many cities, towns, and even civilizations have disappeared. Many of these vanished communities were home to hardworking men and women who were part of the fabric of the nation. When these centers died out, they took some of the country's history with them. *Lost America* takes a look at several locations in America where people thrived and then departed, never to return again.

Although many of these communities have been largely forgotten, they were once vibrant populations with rich cultures. For example, before the United States existed, Cahokia was the largest metropolis in North America north of Mexico. Some historians refer to it as America's first city. In the eleventh and twelfth centuries, as many as twenty thousand people lived in Cahokia. Located near St. Louis, pyramid-like mounds that distinguished this bustling city can still be visited.

The Anasazi of the Southwest were another dynamic people who vanished—but their disappearance has remained an unsolved mystery. Has America also been home to other flourishing kingdoms? Some believe that Vikings established a robust settlement in New England called Norumbega. Others contend that ancient tribes built cities of gold in America that remain hidden to this day. While clues have been found, conclusive proof of their existence has yet to be discovered.

One of the most famous communities that disappeared was the colony of Roanoke. Established in 1585, the settlement was intended to be the start of England's colonization of America. The determined men, women, and children who traveled far from their homeland to live on the coast of North Carolina, however, inexplicably vanished. What happened to these individuals remains one of America's greatest puzzles.

Weather has wiped out several towns and cities. In Louisiana, Texas, Oregon, and elsewhere, hurricanes and floods have washed away thriving communities. In some cases, populations met their demise from man-made causes. Toxins and deadly gases forced citizens in Centralia, Pennsylvania, and Love Canal, New York, to abandon their homes and start life anew in other regions.

Through photographs, drawings, and newspaper clippings, *Lost America* recounts the tales of the towns, cities, and civilizations that are no more, revealing the causes of their demise, unraveling a few of the secrets behind their upheaval, and considering a few of the myths and fringe theories surrounding their disappearance. Although these populations may have ceased to exist, many have had an impact on society, and their struggles and achievements provide invaluable lessons for communities today. ■

PART I

LOST KINGDOMS OF EARLY AMERICA: THE MYTHICAL AND THE REAL

THE HEADLINE

LOCATION: Southern Illinois between East St. Louis and Collinsville, directly across the Mississippi from modern-day St. Louis

DATELINE: AD 1050–1200

CAHOKIA: NORTH AMERICA'S FIRST GREAT CITY

A huge Native American metropolis on the Mississippi River vanished as suddenly as it appeared.

This Mississippi-era coin depicts a common Cahokian design, with a motif of a cross in a circle. Wikimedia Commons, photo by Herb Roe

Few people today know about North America's first great city that developed in a time before Christopher Columbus even set foot on these shores. Spread out over six square miles, the city of Cahokia served as the capital of the Indian nation in Middle America, with its population ranging between ten thousand and twenty thousand. Daily activity in this metropolis centered around more than one hundred tall "pyramids," or flat-topped mounds of packed earth, with houses or temples on top. Because so many mounds dotted this region, St. Louis was originally known as Mound City. The mounds were constructed around a series of vast open plazas, including a grand plaza, which was the size of forty-five football fields. Structures were set up with the heavens in mind, often aligning with the sun or the moon.

Some historians have compared Cahokia in size to the great cities of Europe. At times it was bigger than London, which only had a population of about fifteen thousand people in 1100, according to some estimates. The city became a pilgrimage destination where people would gather to celebrate with games, rituals, ceremonies, and feasting. Some archaeologists have described Cahokia as a Native American Jerusalem. One scientist said it was like the New York City of its day—"the greatest show on Earth."

In the space of 150 years, Cahokia quickly boomed, prospered, and then rapidly disappeared. Designated a UNESCO World Heritage site, about 70 mounds are now protected, and tourists today can visit these remnants of the vast and vibrant cultural center that once was.

Found in a temple near the boundaries of a ceremonial site, the pipestone Birger figurine shows the sophisticated artwork of the Cahokian culture. Wikimedia Commons, photo by Tim Vickers

"It's about time that Native Americans and non-native citizens realize that in the eastern woodlands of the United States, a great civilization arose, and the art it produced is equal to the art of societies at a similar level of development anywhere in the world, at any time and place," says Kent Reilly, a professor of anthropology at Texas State University told *The Christian Science Monitor*.

WHAT DO WE KNOW?

Because the people of Cahokia kept no written records, its history has been largely lost to time. Archaeologists have had to piece together its past from remaining artifacts, bones, and folklore passed down from related tribes. Tribes such

The First Terrace Monks Mound.
Photo by Ben Bridgforth

as Quapaw, Omaha, Pawnee, Chickasaw, Ponca, Mandan, Choctaw, and Osage likely called Cahokia home at one point. Many Native American tribes that trace their heritage to Cahokia believe that it could have been the grand homeland—a central place where all tribes came and lived together before they were differentiated.

Cahokians represent the apex of Mississippian culture. The Mississippians were different groups of Native Americans who thrived from about AD 700 to the time of Columbus, primarily living in the river valleys of what is now Mississippi, Missouri, Alabama, Georgia, Arkansas, Illinois, Indiana, Kentucky, and Ohio. Their cultures shared many similarities. Collectively, they became known as the mound builders because their settlements featured pyramid-like structures or mounds. The communities were laid out symmetrically around plazas with very similar earthworks using the same kinds of construction techniques. Day-to-day activity centered around a major "mound" and a large communal plaza. Generally, a larger, well-populated central area dominated over smaller satellite communities. Their food supplies depended heavily on corn, along with squash and beans and other crops.

The Mississippian cultures displayed sophistication, producing fine pottery, tools, knives, arrowheads, and tattoo kits. They

TIMELINE

500 BC–AD 100	Adena mound builders thrive
200 BC–AD 500	Hopewell people flourish
250–900	Peak of Mayan civilization
900–1168	Toltec culture.
1050	Cahokia established
1150	Fence built around Cahokia
1200	Cahokia declines
1300–1521	Aztec civilization
1925	Cahokia Mounds protected as a public state historic site

Life in Cahokia centered around a vast plaza encircled by pyramid-like mounds. Courtesy of Cahokia Mounds State Historic Site

Monks Mound covers an area larger than the Great Pyramid of Giza. Wikimedia Commons

crafted items from shell, stone, wood, clay, and copper—all intricately molded, carved, and/or engraved. They knew how to mine and smelt metals. Often, designs incorporated a cross, circle, and spiral. The Mississippians depicted the cosmos as a cross within a circle. A cross showed how the world was governed by the four directions. These images often occur in Native American art to this day.

One very detailed piece that has survived largely intact from this era is called the "Birger" figurine. Carved of pipestone (a red, carvable rock that Native Americans often used to make pipes), it depicts a woman in a short wraparound skirt holding a short-handled stone hoe. She's chopping what appears to be the back of a snake that has the head of a cat. The cat's tail transforms into a squash vine as it climbs up the woman's back.

The Early Mound Builders

This mound-building culture in America began prior to the Mississippians with the Adena people in 500 BC, who primarily lived in southern Ohio, but also in West Virginia, Kentucky, Indiana, Pennsylvania, New York, Maryland, and Wisconsin. They built conical burial mounds to honor their tribal leaders. As this culture died out around AD 100, a people known as the Hopewell flourished in Ohio and other parts of Northeastern America. Lasting until about AD 400 or 500, Hopewell culture mirrored that of the Cahokians in many ways. They relied heavily on a corn diet supplemented by other vegetables and wild game. They made ornate pottery and developed a trade network that may have extended as far south as the Gulf Coast, as far east as the Atlantic, and as far west as the Rocky Mountains.

The most famous Hopewell site today is the Newark Earthworks of Ohio—the largest set of geometric earthen enclosures in the world. Considered a sacred site by Native Americans, the area is part cathedral, part cemetery, and part astronomical observatory. Covering four square miles, the Earthworks includes two giant circles, various burial mounds, an ellipse, a square, and an octagon, all connected by parallel walls. Scientists have discovered that the octagon and circle align with the 18.6-year lunar cycle's northernmost moonrise.

With a population as high as twenty thousand, Cahokia has been called "America's first city" by some historians. Courtesy of Cahokia Mounds State Historic Site

Newark Earthworks of Ohio. Wikimedia Commons

THE EVIDENCE: A BIG BANG AND QUICK GROWTH

After the Hopewell civilization faded, the Mississippian tribes formed. Around 1050, Mississippians established Cahokia so quickly and suddenly that it is said to have started with a "big bang." Some researchers speculate that the appearance of a supernova in 1054 may have spurred people to action. Some say a charismatic leader may have inspired thousands of individuals from various tribes to move there within a short amount of time. Ultimately, why Cahokia sprang to life remains a mystery.

Originally, the metropolis probably wasn't named Cahokia; that name came from a subgroup of the Illini Indians who lived nearby in the 1600s and 1700s. Cahokia's influence spread, spawning other related but smaller mound-building communities such as

the Etowah in Georgia, the Spiro people in Oklahoma, and the Mound Builders of Alabama, who erected the second-largest Mississippian city, known as Moundville.

Cahokia sat near the confluence of the Mississippi, Illinois, and Missouri rivers, waterways that served as arteries for travel. Based on materials found here, trade seems to have thrived. Scientists have found copper and mica from the Great Lakes and an engraved whelk shell that originated in the Gulf Coast, more than six hundred miles away. The Cahokian trading network appeared to stretch north to south from upper Michigan down to the Gulf of Mexico, and east to west from the Smoky Mountains to the Ozarks. Communities naturally gathered around rivers because they fed fertile lowlands. In addition to corn, Cahokians grew goosefoot (also known as wild spinach) and the grain amaranth.

The Amazing Monks Mound

Most of the population lived in smaller homes with thatched roofs in the shadow of Monks Mound. (named after the Trappist monks who settled in the region several hundred years after it was built). Monks Mound rose up about one hundred feet (ten stories tall), with a base that covered about fourteen acres, slightly bigger than the base of the Great Pyramid of Giza in Egypt, making it one of the largest pyramids ever built. Like the ancient Egyptians, Cahokians appear to have built this tall structure to get closer to the heavens. Plus, the rulers who lived atop Monks Mound could survey their people and lands and show their dominance. Today, when the weather is clear, visitors to the mound can see the St. Louis Arch from its apex.

Most of the people lived in smaller homes with thatched roofs in the shadow of Monks Mound. Courtesy of Cahokia Mounds State Historic Site

Corn was a major part of the Cahokian diet. While it provided sustenance, eating too much corn may have led to poor nutrition among these Native Americans. Courtesy of Cahokia Mounds State Historic Site

A circle of posts (or henge) formed a great sundial observatory. If you stood at its center on the mornings of the spring and fall equinoxes (usually March 21 and September 21), you would see the sun rise from behind a post and directly out of Monks Mound. Around the winter solstice (December 21), the sun would be seen to rise above four posts to the south, and around the summer solstice (June 21), the sun rose above four posts to the north. (Over time, at least five different woodhenges were built at this location.)

The Cahokians likely held ceremonies connecting the resurrection of the dead with the renewal of the earth. Members of the Choctaw tribe still perform ancient spiral snake dances at sunrise at this location, celebrating a connection with nature as their Cahokian ancestors did.

THE ROLES OF MEN AND WOMEN

ircling the pyramids were thousands of pole-and-thatch homes, temples, and public buildings, according to Timothy Pauketat's book, *Cahokia: Ancient America's Great City on the Mississippi.* Beyond those lay a fifty-mile ring of towns and farms where crops were raised to feed the swelling populace.

In 1995, ten miles southeast of central Cahokia on the edge of the Illinois prairie, archaeologists uncovered remnants of about 150 houses and sheds. The excavation team found no male-related artifacts, such as arrowheads, but did discover bone weaving tools, spindle whorls (for spinning fibers into yarn), and evidence of farming, cooking, and pot making. Testing of the remains revealed that people here ate a diet of small lizards, snakes, rodents, and turtles, along with huge amounts of corn. Scientists believe that this was a peasant population who worked to feed and supply the elite in the heart of Cahokia.

In this diorama from the Cahokia Museum, a woman grinds maize using a mortar and pestle. Wikimedia Commons, photo by Herb Roe

Some historians speculate that women may have had influence—at least for a time—not just in looking after the fields, but also in controlling the inheritance of land. Still, evidence points to a more male-dominated culture. While women dressed simply, high-ranking men had more exotic hairstyles and wore ear spools.

In Cahokian art, men are depicted hunting or playing games with balls, rings, hoops, and pins. The most popular game was called "chunkey." Also called *chenco* or *tchung-kee*, the name roughly translates to "running hard labor," according to ethnographer James Adair. Men rolled round discs called "chunkey stones" across the flat plazas. They would bet on where they would stop, or throw spears (estimated to be eight or nine feet long) to halt them. Players competed to hit them with javelins. Points were earned according to how close to the stones one could get the sticks. Chunkey may have been a later version of a children's game called hoop and pole. One's standing in the community was affected by his skill as a chunkey player. Some took losing very hard and even committed suicide as a result.

George Catlin painting of Mandan Indians playing Chunkey in 1832, showing the game reached into historic times. Wikipedia Commons

THE DARK SIDE OF CAHOKIA

While art, architecture, crafts, and farming flourished, the civilization had a more violent side as well. At Mound 72, a mile south of Monks Mound, archaeologists found two male corpses. One lay under a two-inch-thick layer of twenty thousand shell beads. Shells were rare, and the sheer volume indicated that these men were important members of society. The other corpse lay on top of the beads. The beads appeared to have been sewn to form some type of blanket or cape, but the fabric had disintegrated to dust. The beads appeared to form the shape of a falcon or thunderbird. Wings, tail, and head paralleled the man's body. In nearby pits were hundreds of arrowheads, thirty-six thousand beads, two bags full of chunkey stones, precious copper, and mica.

As the scientists dug nearby, they uncovered the skeletons of thirty-nine men and women who had met a violent death. They

A mass grave with the bodies of fifty-three women revealed that the Cahokians may have practiced ritual sacrifice. Wikimedia Commons, photo by Herb Roe

had been lined up along the edge of a pit and hit with stone axes, dropping them dead into the ditch. Scientists could tell that some were still alive when buried because their finger bones had dug down into the clay.

Like many other ancient civilizations, the Cahokians shared a belief in the power of sacrifice. Archaeologists uncovered several examples where sacrificial victims were buried beneath temple mounds. One pit contained the skeletons of four headless and handless men. Another mass grave held the skeletons of fifty-three young women between the ages of eighteen and twenty-five. Native tribes at the time were known to sacrifice individuals who might act as servants in the afterlife. Another type of sacrifice was an offering to the gods—killing a person as part of a ritual to ensure a good harvest of corn. Their deaths were thought to contribute to the life of the natural world and to benefit the earth.

DEVELOPING STORY:
WHEN A MIGHTY NATION FALLS

A sign of trouble in Cahokia came around AD 1150, one hundred years after the city began. At that time, the people built a two-mile-long stockade of vertical logs around the metropolis for protection. Why did a mighty nation suddenly need to build a wall? Did the rulers of Cahokia perceive some sort of threat that might topple their reign? No one is certain of the answer, but historians say that when the wall went up, it was the beginning of the end of a once-powerful nation.

Encompassing two hundred acres and eighteen mounds, the fence had twenty thousand poles, requiring an amount of wood that would have severely depleted a valuable natural resource.

The Cahokians used wood for building, cooking, and heating. Evidence found that Cahokians switched from using long-burning oak and hickory in their early days to less-efficient softer woods near the end—a sign that all was not well. Archaeologists think deforestation had a devastating effect on their agriculture and survival. The lack of trees may have driven deer from the area, taking away from the food supply.

If the protein supply dwindled, diet may have played a role in the decline as well. The people survived largely on an abundance of corn, but eating mostly maize did not provide a very balanced diet, leading to an iron deficiency. Since there were no signs of garbage or human waste disposal in Cahokia, dysentery and tuberculosis due to poor sanitary conditions may have contributed to their demise as well.

Natural disaster may have added to the woes in Cahokia. Botanists and geologists who have analyzed tree rings and soils identified a period of multiple droughts during the span between AD 1150 and 1300. On the other hand, flooding could have occurred as well. A global cooling trend called the "Little Ice Age" did occur around 1250, which may have hurt the growing season. But by this time, Cahokia is thought to have already collapsed. Either way these disasters could have wiped out crops needed to survive. Pastures to the west may have looked greener, where herds of buffalo offered an abundant supply of red meat.

Signs of Turmoil

Although there may have been some outside threat to Cahokia that prompted the building of a wall, historians also believe turmoil may have come from within. Power struggles may have broken out among the clans that had once operated in harmony. Some experts believe that women lost their control of the system of land inheritance, and a shift occurred toward an even more male-dominated society. Corn goddess figurines were smashed or burned. In the later period around 1200, more images appeared featuring the falcon or thunderbird with their threatening link to power, death, and conflict. Red clay Cahokian statues show a naked warrior crouching behind a wooden shield, and another shows a Cahokian in full battle gear, wearing thick cloth armor to protect his body from enemy blows; a shield is hung around his back, and he is leaning over a fallen foe, clutching his hair.

As dissatisfaction increased among the people, the leaders built a lower terrace around Monks Mound so they could be more visible to the population. The theory is that more visibility would help to keep the trust of the people. As the empire began to crumble, so did the mound itself. Around 1200, about one-quarter of the mound fell off in a landslide. Scientists note that an earthquake may have been the cause.

By 1200, a civilization that had seen more than one hundred years of peace and prosperity was coming to a violent end. Archaeologist Timothy Pauketet writes that the people may have grown tired of living under the control of one empire. Perhaps the Cahokian rulers "pushed the envelope of authority, taxation, and oppression too far." The Cahokians splintered off into tribal groups in search of better lives. The unified nation may have rebounded to its former glory, but in the 1500s the greatest threat to Native American people came as explorers and settlers arrived from Europe.

LASTING IMPACT: PRESERVING NATIVE AMERICAN HISTORY

Early Americans were well aware of the mound-building culture. In the early 1800s, Lewis and Clark reported their discovery of the mounds to Thomas Jefferson. In the nineteenth century, as Americans settled farther west, many of the mounds were plowed over and destroyed. Discoveries of Cahokian antiquities conflicted with the notion that Europeans were superior and that Native Americans were savages. Many white settlers supported a theory called "diffusionism," attributing the architecture, art, science, and achievements of Cahokia to another culture. American Indians simply could not have developed such a sophisticated culture. Maintaining this view that Native inhabitants were inferior provided justification for taking their lands. Archaeological discoveries in the 20th century about Cahokia and other mound-building cultures have led to efforts to preserve the remaining structures, artifacts, and record the true history of this remarkable culture.

As early as the 1800s, American settlers were aware of the mounds, but unsure of their origins. Many, like this one photographed in Illinois in 1907, became part of thelandscape.
Library of Congress

FRINGE THEORIES

1 A Connection to Mexican Civilizations?

Although no absolute proof has been found, the parallels between Cahokia and Mexican cultures like the Mayans and Aztecs are undeniable. The Mayan culture of Central America and Mexico flourished between about 2000 BC and AD 900 AD. The Aztec empire thrived from 1300 to 1521, after Cahokia had already disappeared. The pyramid-shaped mounds and vast plazas of Cahokia echo the layouts of Mayan and Aztec cities. Some scientists note that Monks Mound closely resembles the Great Pyramid of Cholula near the city of Puebla, Mexico, which was completed in the eighth century.

As the Mayan civilization faded, the Toltecs rose to become a dominant culture in the tenth to twelfth centuries, a time period that overlapped with the rise of Cahokia. The Toltec capital of Tollan, north of Mexico City, had a plaza similar in shape and scale to that of Cahokia. At its height, Tollan had about fifty thousand inhabitants. (Note that the Toltec Mounds in Arkansas were part of the Mississippian culture. They were named *Toltec* because of their resemblance to that culture, but have nothing to do with Mexico.) The Aztec pyramids and plazas in the heart of Tenochtitlan in Mexico greatly mirror Cahokia as well.

Culturally, these Mexican cultures also practiced human sacrifice and had a great interest in astronomy. A recurring motif

among Cahokian art is the Long-Nosed God, a figure with a very long, Pinocchio-like nose. The image has been etched in shells, clay, and copper. Some researchers have noted the resemblance of this figure to images in Mexican culture, such as the god Quetzalcoatl, who was revered by the Aztecs. Anthropologists observe that corn and squash, which were first domesticated in Mexico, somehow made their way north and became the dominant items in the Cahokian diet, as well.

Scientists have also found similarities between the cultures of the Creek, Seminole, Miccosukee, and Chitimacha Indians of the the Southeast and the Mexican cultures of Mesoamerica.

Aspects of Mexico's Toltec society mirrored Cahokia, including the flat-topped pyramids that resembled Cahokia's mounds. Wikimedia Commons, photo by Susana Torres Sánchez

2 A Connection to Ancient Hebrews?

Some researchers maintain that the mound builders and other Native Americans share a connection to ancient Jewish travelers. In June 1860, David Wyrick and his colleagues were exploring a mound ten miles south of Ohio's Newark Earthworks. They found a wedge-shaped stone with inscriptions in Hebrew that read *Melech Eretz* ("King of the Earth"); *Toras Hashem* ("The Law of the Lord"); *D'var Hashem* ("The Word of the Lord"), and *Kodesh Kodashim* ("Holy of Holies"). A few months later, Wyrick made a more shocking discovery: a black stone inscribed with the Ten Commandments in an ancient form of Hebrew. Wyrick said that these artifacts, known as the Newark Holy Stones, proved that the mound builders were connected to the Lost Tribes of Israel. Mainstream archaeologists, however, accused Wyrick of faking the stones and staging an elaborate hoax, noting that they did not appear to be old enough.

In 1889, researchers with the Smithsonian Institution excavated a burial mound in eastern Tennessee and uncovered what became known as the Bat Creek Stone. At the time, the project director declared that the inscription was a form of the Cherokee language. In the 1960s, however, two Semitic scholars noticed that when they turned the stone upside down, the characters appeared to be an ancient form of Hebrew. Was this additional proof that Israelites had wandered the globe, making it as far as North America? In 2004, two researchers published evidence that debunked this theory. The inscription was copied from an illustration that appeared in a widely available book titled *General History, Cyclopedia, and Dictionary of Freemasonry*, published in 1872.

Some scholars believe that genetic evidence also indicates a connection between the mound builders and the Israelites. DNA evidence extracted from the remains of Hopewell cemetery pop-

ulations revealed some DNA lineage linking the Hopewell people to ancestors from the "hills of Galilee" in Israel. Some researchers believe that genetic evidence shows there was a great migration tens of thousands of years ago, across the Bering Strait into North America. The migrating people included those who originated in the Middle East.

Originally thought to be Cherokee, an inscription on the Bat Creek Stone was later discovered to be an ancient form of Hebrew. Was it a hoax, or evidence that the Israelites had traveled to North America? Wikimedia Commons, photo by Scott Wolter

THE
HEADLINE

QUIVIRA AND THE SEVEN GOLDEN CITIES OF CIBOLA

Lost treasure city turns out to be merely a kingdom of hay houses

LOCATION:
Great Bend, Kansas

DATELINE:
Late 1400s–1500s

Beginning in 1492 with the arrival of Christopher Columbus in the Caribbean, the Spanish conquered a vast amount of territory in the Western Hemisphere, including the Caribbean Islands, most of Central America, half of South America, and the southern and western portions of North America. In 1521, Hernán Cortés and his conquistadors leveled the capital city of the Aztecs in Mexico, putting an end to the once-mighty empire. Twelve years later Francisco Pizarro and his army wiped out three hundred years of Incan civilization in Peru when they executed the last Incan emperor.

In 1538, when Francisco Vázquez de Coronado served as governor of the province of Nueva Galicia, an area in northwestern Mexico, he heard rumors of a Mexican kingdom filled with gold that would rival the Aztec empire. In 1539, a Franciscan monk, Friar Marcos de Niza, told Spanish officials in Mexico City that he had seen the legendary Seven Golden Cities of Cibola located to the north, where the present-day Native American pueblo of Zuni, New Mexico, is located. The problem is, he only saw Cibola from a distance; because his traveling companion had died, he decided to turn back before exploring further. Still, he described this distant metropolis as one of the most beautiful kingdoms he had ever seen.

In 1540, with the prospect of vast riches lying to the north, Coronado mounted an expedition of three hundred Spaniards and more than one thousand Indians. To sustain themselves, they readied hundreds of horses, cattle, sheep, and pigs as well, and then set off with high hopes of finding the famous cities of gold.

A kiva, such as the one pictured here at Gran Quivira, was an area used by Puebloans for religious rituals and political meetings. Wikimedia Commons, photo by HJPD

Today part of New Mexico's Salinas Pueblo National Monument, Grand Quivara was a vast city with multiple pueblos that Spanish conquistadors found during their search for a city of gold. Wikimedia Commons, National Park Service

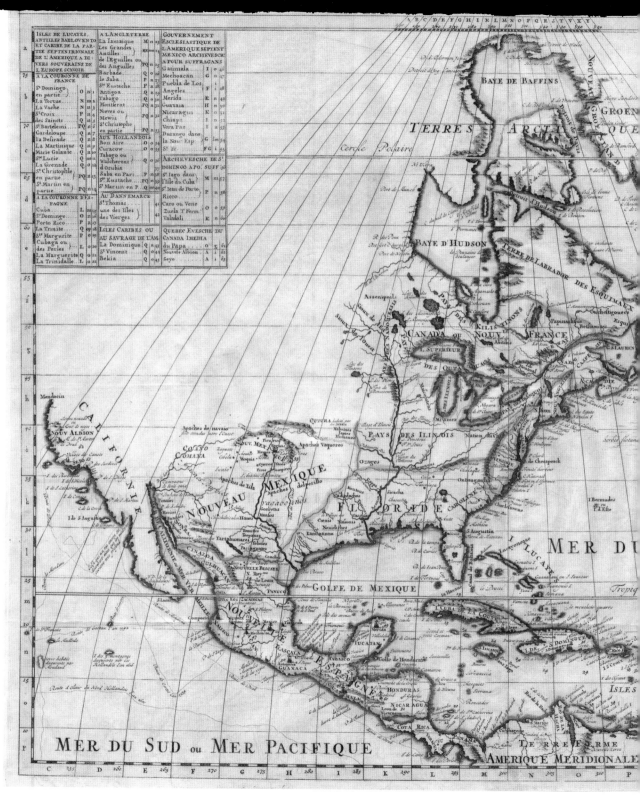

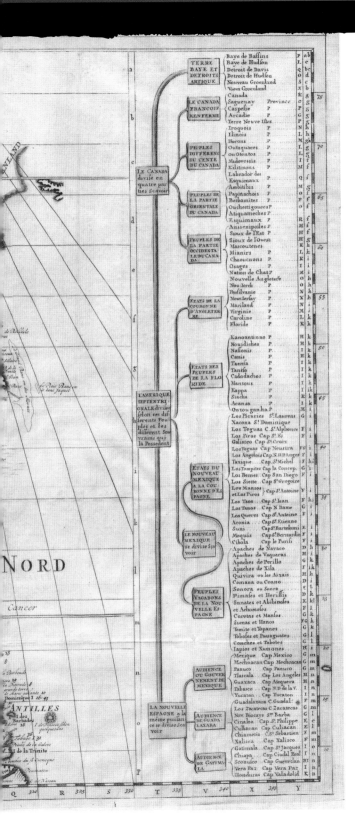

Map of North America published by Henry Chatelain for the 1720 edition of his *Atlas Historique*. It includes references to both Quivra (Quivira) just west of the Mississippi, and Cibola in New Mexico, two of the "Seven Cities of Gold" sought after by Coronado. Both turned out to be modest Indian villages, much to Coronado's disappointment.
National Park Service

WHAT DO WE KNOW?

Although Coronado spent two years intensively searching for Quivira, his expedition ended in failure and ultimately led to his downfall. Courtesy of Billy Hathorn, Deaf Smith County Museum in Hereford, Texas

In February 1540, Coronado and his huge entourage embarked northward through the western part of the continent, heading to regions that would become the southwestern United States. As part of the mission, another conquistador, Hernando de Alarcon, packed two ships on the Pacific Coast of Mexico and headed north with supplies in May.

Although Alarcon was to meet up with Coronado's land expedition at an appointed latitude along the Pacific Coast, he never connected with Coronado's army. He did, however, find the mouth of the Colorado River, which empties into the Gulf of California on the Pacific Ocean near Baja California and Sonora in Mexico. Alarcon buried a message stating that he had waited, but then eventually gave up and continued on. A short time later, Coronado's men found the message, but missed out on getting the supplies they desperately needed. Alarcon meanwhile explored farther up the Colorado River. One of his crew members on that expedition drew a map that is considered one of the earliest and most accurate representations of the lower Colorado River and the Gulf of California. One of the men dubbed the river Firebrand, a name it maintained for two centuries.

In the summer of 1540, Coronado and his men arrived in what is today New Mexico, at Hawikuh, one of the largest of the Zuni Indian pueblos at the time, and also rumored by some to be one of the famous seven cities of gold. The Indians here battled the conquistadors the best they could, but they were no match for Coronado's army. After six weeks of occupying this Indian community of earth and stone, Coronado wrote, "As far as I can judge, it does not appear to me that there is any hope of getting gold or silver, but I trust in God that, if there is any, we shall get our share of it." So far, however, Coronado's expedition had been a bust.

Toward the end of August, two Indians arrived in Coronado's camp. Coronado showed them great hospitality, and in return, they told the Spanish explorer of a magnificent pueblo to the east. He also learned of huge numbers of "cattle" (bison) that roamed free on the plains. Intrigued by these tales, Coronado sent his trusted captain Hernando de Alvarado to explore these pueblos to the east, taking along about twenty conquistadors and a friar. They discovered the Acoma Pueblo and Rio Grande Pueblo province, home of the Tiguex villages, all near today's Albuquerque. While the pueblos were impressive communities, they were not overflowing with riches.

Leaving a contingent of men at Hawikuh, Coronado left with another group to spend the winter of 1540–41 in the Tiguex villages. In his typical ungracious manner, Coronado invited the Native American residents in Tiguex to leave and find other accommodations.

Without the supplies they had counted on from the ships, Coronado and his men were desperate for resources during the winter. The Spanish soldiers had been motivated to find gold, silver, and riches, but they had become increasingly restless. Coronado grew uneasy as well; he had spent a small fortune on a quest that so far had been fruitless.

In addition, the Native Americans in many of these pueblos had grown resentful and angry. Many fought bloody battles against Coronado's forces, but the powerful Spanish forces slaughtered hundreds. To demonstrate their supremacy, they often burned entire villages to the ground. The Spanish regarded themselves and their God as superior: If the Indians did not bow down to them and follow their religion, the Spaniards would kill them off.

One Indian who served as a guide to the Spanish gave them hope. He told tales of a province called "Quivira," one of the fabled kingdoms that overflowed with gold and silver. He described "trees hung with golden bells and people whose pots and pans were beaten gold." Nicknamed "the Turk" (because to the

These two actors are dressed in the style of the Spanish conquistadors who searched the Southwest for the lost city of gold. Wikimedia Commons, public domain

Spanish, he resembled a Turkish man), the Indian guide said he would take them to this land of immeasurable wealth. Buoyed by the Turk's good news, Coronado and his men forged ahead with their mission.

THE EVIDENCE: A MASSIVE EXPEDITION EXPLORES AMERICA

According to historian Herbert Eugene Bolton, in April of 1541, Coronado renewed his quest to find Quivira, departing with more than fifteen hundred conquistadors, Indian aides, servants, and slaves. Coronado and his entourage were not into traveling light—they took along one thousand horses, five hundred cattle, and some five thousand sheep.

The group likely traveled north along the Rio Grande, turning northeast at the end of the Sandia Mountains, which lie just east of Albuquerque. Then they marched into the vast flat Great Plains. Toward June, a discouraged Coronado sent much of his army and people back to the pueblos of Tiguex. A smaller force accompanied Coronado north into Kansas, where they believed Quivira awaited.

When Coronado finally arrived in Quivira that summer, it turned out to be nothing more than a village of grass houses near what is now Great Bend, Kansas. They found some iron pyrite and copper, but not an iota of gold or silver. According to expedition accounts, many of Coronado's men were so disappointed they threw down their armor in anger and disappointment. Realizing that the Turk had deceived him, Coronado had the Indian strangled. Coronado and his group stayed with the natives of Quivira approximately three weeks and then headed back toward the Tiguex villages.

On the return trip, Coronado reportedly slipped from his horse and hit his head. He became despondent about his injury, the failure of his mission, and being apart from his wife.

TIMELINE

1492   Christopher Columbus reaches the New World

1493   Ponce de Léon accompanies Columbus on his second expedition to the Americas

1513   De Léon discovers Florida

1540   Francisco Vázquez de Coronado sets out from Mexico to find the Seven Golden Cities of Cibola

1541   Coronado discovers Quivira, which is a large Indian settlement of hay homes in Kansas, but no riches

1542   Coronado leads his forces back to Mexico after the failed expedition

DEVELOPING STORY :
THE JOURNEY ENDS IN DISAPPOINTMENT

In October of 1541, shortly after arriving back in the pueblos, Coronado put the best face on the expedition and wrote the King of Spain:

> The province of Quivira is 950 leagues from Mexico. Where I reached it is in the 40th degree. The country itself is the best I have ever seen for producing all the products of Spain, for besides the land itself being very fat and black and being well watered by the rivulets and springs and rivers, I found prunes like those of Spain, and nuts, and very good sweet grapes and mulberries. I had been told that the houses

were made of stone and were several storied; they are only of straw, and the inhabitants are as savage as any that I have seen.

They have no clothes, nor cotton to make them of; they simply tan the hides of the cows which they hunt, and which pasture around their village and in the neighborhood of a large river. They eat their meat raw like the Querechos and Tejas, and are enemies to one another and war among one another. All these men look alike. The inhabitants of Quivira are the best of hunters and they plant maize.

Coronado and his men may have not found gold but they did discover many natural riches of the Great Plains, such as herds of bison. Wikimedia Commons, public domain

By the spring of 1542, Coronado led his men out of the region and back to Mexico. They wanted the quick rewards of finding gold and silver but did not see the long-term gains of developing these rich agricultural properties. Coronado trudged back as a failure. He resigned his commission as governor of New Galicia and disappeared from the public eye. Not only did he not find gold, but his brutal tactics in North America also established distrust and hatred among the Indians. Instead of forging a peaceful relationship, many Indians would see the Spanish and Mexicans as their enemies for centuries to come

LASTING IMPACT : THE LEGEND OF A GOLDEN CITY LIVES ON

The dream of finding Quivira did not die easily. Many believed that there was still a City of Gold somewhere in America. In 1601, Spanish explorer Juan de Onate organized an expedition, determined to find the real Quivira. Onate took a small group of approximately one hundred soldiers back into the area that is now Kansas and discovered the city of Etzanoa, populated by about twelve thousand to twenty thousand people. Unlike other Native settlements like Cahokia, there were no large public spaces or pyramid-shaped mounds, just simple beehive-shaped homes of wood and straw.

Etzanoa, which means "The Great Settlement," was located at the confluence of the Arkansas and Walnut rivers, about 175 miles southeast of Quivira. But again, no immense treasure trove of gold and silver revealed itself. It wasn't until 2017 that modern-day archaeologists located Etzanoa. The year before, high school student Adam Ziegler found a small iron cannonball in the area. Based on this evidence, scientists discovered other artifacts indicating that a massive town once stretching thousands of acres once existed.

Today, Lyons, Kansas, is home to the Coronado Quivira Museum, devoted to Coronado's expedition and the Quivira Indians of central Kansas. The Quivirans themselves evolved into the Wichita tribe. ∎

FAST FACT

At the same time that Coronado searched for Quivira, Hernando De Soto was exploring Florida on behalf of Spain. He had crossed the Mississippi in the spring of 1542, and although they came within hundreds of miles of each other, they never met up.

THE FOUNTAIN OF YOUTH

When it comes to mythical lost locations in America, one of the most famous is the Fountain of Youth. Legend has it that the Spanish explorer Ponce de Léon set off in the early 1500s searching for magical waters that could reverse the aging process and cure ills. De Léon likely began his exploring life as part of Christopher Columbus's second expedition to the Americas in 1493. Like almost every traveler to the New World at the time, he heard the yarns of gold and silver cities ready to be plundered. In 1508–09, when he headed to Puerto Rico to find such treasures, De Léon established what would be the oldest colony on the island, and he became Puerto Rico's first governor.

For decades, Florida has relied on the legend of rejuvenating waters to attract tourists, but no one has yet reversed the aging process. Wikimedia Commons, Boston Public Library, Tichnor Brothers Collection

Pushing on with his explorations, De Léon and his crew set foot in Florida on April 2, 1513—the first Europeans to do so. Because they discovered this land at Easter, De Léon named the location "Florida." *Pascua Florida* is a Spanish term that means "flowery festival," and refers to the Easter season. Although little or no evidence exists that truly shows De Léon was hunting for the Fountain of Youth, he may have heard the traditional stories told by the Taino Indians of the Caribbean about a rejuvenating spring somewhere north of Cuba. The records of the time show that De Léon was more interested in colonizing, but tales grew that he was obsessed with finding the miracle waters that could make people young again. Far from finding a source of eternal life, De Léon proved he was mortal. After being shot in the leg with an Indian arrow, his wound festered and killed him a short time later in 1521.

Over time, the state of Florida embraced the legend; being known as the land of eternal youth is good for business. In the early 1900s, a statue of Ponce de Léon was erected in the center of St. Augustine. Visitors to the city can learn about the first Spanish settlers at the Fountain of Youth Archaeological Park. Tens of thousands of tourists now sample the on-site sulfurous well water, which has been dubbed the "fountain of youth," but smells more like the "fountain of old rotting eggs." St. Augustine isn't the only Florida city laying claim to the rejuvenating waters. Punta Gorda, about 250 miles southwest, has its own foul-smelling spigot of youth, which is about the size of a garbage can. The public can drink freely from this source, but it does display a warning sign from the Florida Department of Health: USE WATER AT YOUR OWN RISK. The water from this well exceeds the maximum contaminant levels for radioactivity as determined by the US Environmental Protection Agency under the Safe Drinking Water Act. Visitors still gulp it down. Apparently the dream of reversing the aging process beats the possibility of dropping dead from radioactive contaminants. If neither of these water sources induces youth, Florida offers a few other locations scattered about the state that claim to be the "original" Fountain of Youth.

From the time Ponce de Léon explored Florida in the early 1500s, the Fountain of Youth has captured the imagination of all those hoping to be forever young. Library of Congress, photo by John Margolies

Many sites in Florida claim to be the real Fountain of Youth, like this one at the Archaeological Park in St. Augustin. But most look more likely to shorten lives than extend them. Diego Delso, delso.photo, License CC-BY-SA

THE HEADLINE

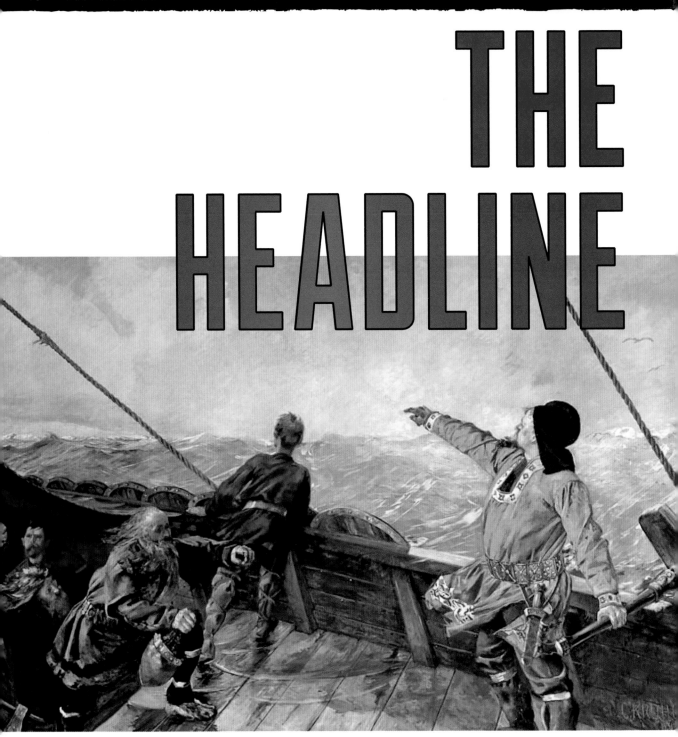

While Leif Eriksson, pictured here, made it to Newfoundland in his expeditions, some believe that he may have explored New England as well. Painting of Leif Eriksson discovering America by Christian Krohg, National Gallery of Norway

NORUMBEGA: THE VIKING CITY OF NEW ENGLAND

LOCATION:
New England

DATELINE:
AD 1000–?

Was the Northeast once the heart of a striking Viking empire established by Leif Eriksson?

In 1568, when the English sailor David Ingram was rescued after having been lost in America for eleven months, he returned to England with some incredible tales to tell, including a detailed description of the mythic Viking kingdom called Norumbega. Although inhabited by Native Americans, the city that he reportedly stumbled onto had been established by the Vikings and sparkled with vast amounts of gold, silver, and jewels. He told any who would listen that the residents there enjoyed an abundance of food, furs, and other wealth. His stories fed the imaginations of English adventurers such as Sir Walter Raleigh, who hoped to discover such vast riches in the New World. For decades, mapmakers had labeled a sprawling region in New England as "Norumbega." Did an opulent Viking metropolis by this name actually exist?

Erected by Eben Norton Horsford in 1889, Norumbega Tower supposedly marked the location of Fort Norumbega in the legendary Norse city in North America. Library of Congress

WHAT DO WE KNOW?

O n several maps from the 1500s, the name Norumbega clearly marks much of the land that is now New England in the United States. Giovanni da Verrazzano was thought to be the first European to see New York Bay. His brother, Girolamo, labeled the New England region "Oranbega" on a map he created in 1529 based on Giovanni's reports. Some historians believe the name comes from the Italian *non oro bega*, meaning "no gold to quarrel about." Italian cartographers labeled the region as such on maps to signal that New England was in fact an area void of wealth. Other scholars believe the Algonquin Indians first came up with the name, which translates to "quiet place between the rapids" or "quiet stretch of water." Some believe, however, that the American Indians derived "Norumbega" from "Norvega," meaning Norway. So perhaps Scandinavian explorers had been in the region after all.

About five hundred years before Columbus arrived in North America, Viking raider Leif Eriksson and a crew of thirty Norse sailors likely explored the Maine coast, hoping to establish a settlement there. According to one theory, Eriksson blew off course on a trip to Greenland around AD 1000 and landed on the North American continent in a region he dubbed Vinland ("Land of Wine"). In 1960, archaeologists unearthed evidence of a Norse community at the northern tip of Newfoundland in L'Anse aux Meadows, believed to be Vinland. No more than 130 people had settled there, according to investigators.

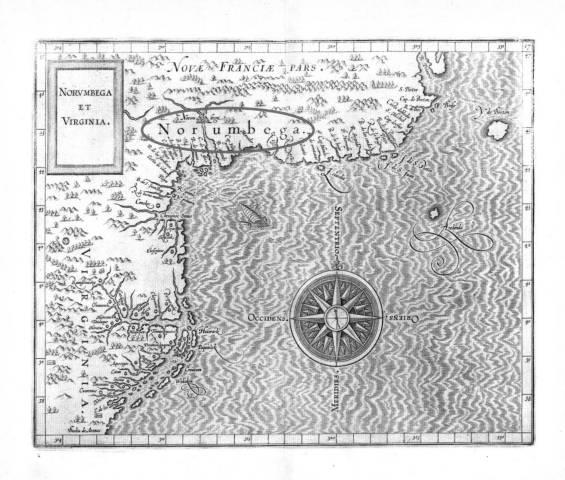

Many old maps clearly labeled a large part of New England as Norumbega. Courtesy of Norman B. Leventhal Map Center, author: Corneille Wytfliet

THE EVIDENCE:
A KINGDOM BASED ON ONE MAN'S WORDS?

T he most compelling evidence that the Vikings had built a paradise of riches in New England came from a mariner, David Ingram, who had an uncanny gift for weaving a story.

In November of 1567, Ingram found himself and one hundred shipmates marooned near Tampico on the coast of Mexico, about two hundred miles south of the present Texas–Mexico border. To avoid capture by the Spanish, he and his crew struck out by foot in a northern direction and disappeared. About eleven months later, in October 1568, a French fishing vessel off the coast of Nova Scotia crossed paths with Ingram and two of his shipmates. They hauled them aboard and agreed to sail them back to England.

Back in England, Ingram told his tale to any who would listen. Ingram said that he and his men had walked the entire distance from Mexico to Canada in less than a year. Although many said it would have been physically impossible to walk over three thousand miles through the wilderness in only eleven months, the British writer Richard Nathan retraced Ingram's journey in reverse in 1999, walking from Nova Scotia to Tampico. He completed the trek in just nine months.

Ingram's trip by foot was adventure-filled. He described wandering into the opulent land of the ancient Vikings. The mythic city centered around what is today Bangor, Maine, and along the Penobscot River. He told of Native Americans who wore jewelry inset with thumb-size pearls. They resided in round houses held up by pillars of crystal, gold, and silver.

In one account that he told a transcriber wrote: "He saw kings decorated with rubies six inches long; and they were borne on chairs of silver and crystal, adorned with precious stones. He saw pearls as common as pebbles, and the natives were laden down by their ornaments of gold and silver. The city of Bega was three-quarters of a mile long and had many streets wider than those of London."

> ## "He saw kings decorated with rubies six inches long; and they were borne on chairs of silver and crystal."

In a great hall, the finest gold lined the walls to the ceiling, which was constructed of silver. Underfoot, Ingram walked across rugs crafted of the choicest skins and furs. The houses along the main street of the city were white and shining, some with roofs of silver and some of copper, with wonderful entrances of crystal, hooded with beaten silver and with doors of burnished copper. The Indian chief, or Bashaba, lived in such a house and bestowed upon Ingram a squaw to cook his food, rich skins to replace his tattered shreds of clothing, and a supply of bows and arrows. The Bashaba invited him to stay as long as he liked.

Ingram described red sheep and other amazing creatures, including one with floppy ears like a bloodhound (perhaps a moose?).

In part of a sworn statement, Ingram said, "The Kings wear great precious stones which commonly are rubies, being four inches long and two inches broad. All the people generally wear bracelets as big as a man's finger upon each of their arms, and the like on each of their ankles, whereof one commonly is gold and two silver, and many of the women also do wear great plates of gold covering their bodies and many bracelets and chains of great pearls."

Sir Walter Raleigh's half-brother, the explorer Sir Humphrey Gilbert, was captivated by these stories. He led colonizing expeditions to North America on behalf of Queen Elizabeth with dreams of finding Norumbega. In 1583, Gilbert arrived in St. John's, Newfoundland, and claimed the territory for England, but soon after he was lost at sea, never to be seen again.

Sir Walter Raleigh and other adventurers believed in Ingram's vision as well. They thought that some sort of North American El Dorado, a city of gold, remained hidden in the northeastern part of the New World. In 1605, the French explorer Samuel de Champlain came looking and reported no evidence of Norumbega. The dream of an opulent Viking kingdom, however, lived on.

TIMELINE

AD 1000	Leif Eriksson discovers Vinland (Canada); possibly explores the coast of New England
c. 1004	Leif Eriksson's brother, Thorvald, dies in North America, possibly in Massachusetts
1492	Christopher Columbus claims the West Indies for Spain
1497	English explorer John Cabot explores Newfoundland
1583	Sir Humphrey Gilbert claims Newfoundland for England
1585	Roanoke settlement established in North Carolina
1603	The French explorer Samuel de Champlain makes first voyage to North America
1889	Eben Norton Horsford announces that he has discovered Norumbega
1957	Eleventh-century Norse coin uncovered in Brooklin, Maine

DEVELOPING STORY: THE VIKING NARRATIVE PERSISTS

I n 1837, Danish historian Carl Christian Rafn published *Antiquitates Americanæ*, the first modern translation of the Sagas (accounts of Viking history), including exploration into North America. In this work, Rafn concluded that Vinland encompassed the coasts of Massachusetts and Rhode Island, as well as settlements on the Charles River near Boston and Mount Hope Bay near Providence. He wrote that Thorvald, the brother of Leif Eriksson, was probably killed and buried around AD 1004 near either Point Allerton in Hull, Massachusetts (on Boston Harbor), or The Gurnet at the entrance to Plymouth Bay. Still, conclusive evidence has not confirmed these events.

One of the few verifiable artifacts has been an eleventh-century Norse silver coin uncovered in Brooklin, Maine, along

Author Henry Wadsworth Longfellow helped to promote the notion that Vikings populated New England with his poem, "The Skeleton in Armor." Uncredited, Public Domain

One of the biggest champions of Norumbega was the inventor of double-acting baking powder, Eben Norton Horsford, pictured here circa 1860. Author unknown, Public Domain

with a few other artifacts excavated at a former Native American trading center. The coin has become known as the Maine penny, and dates back to the reign of Olaf Kyrre, king of Norway (AD 1067–1093). The penny, however, may have been used in trade from Newfoundland, or may have been brought to North America much later by the English or Portuguese.

Still, in the 1800s, prominent New Englanders embraced the notion that mighty Norsemen once ruled Massachusetts, Rhode Island, New Hampshire, and Maine. One believer was the poet Henry Wadsworth Longfellow of Portland, Maine. In 1832, a skeleton wearing a plate of bronze armor was unearthed in Fall River, Massachusetts. While the remains could have easily been those of a Native American or early British colonist, Longfellow and others bought into a popular premise that this was a Viking skeleton— although Vikings never wore bronze armor. Longfellow so liked this possibility that he wrote a poem titled "The Skeleton in Armor." This poem and others fueled further speculation that Vikings had had a presence in New England.

One of the biggest proponents of Norumbega was Eben Norton Horsford. As the inventor of double-acting baking powder, Horsford acquired great wealth and could devote much of his time to his passion—studying the Viking settlement of America. In 1886, he published a work finding that Salem's Neck, Massachusetts, was the site of the English explorer John Cabot's landfall, and that Norumbega was located on the Charles River in the present town of Weston, Massachusetts.

Horsford himself claimed to have found the site of Leif Eriksson's house at Gerry's Landing on the Charles River. Placing a plaque there (which remains today), he declared that the site was a portion of the famous Norumbega Viking settlement. He stated on November 21, 1889: "I have today the honor of announcing the discovery of Vinland, including the Landfall of Leif Eriksson and the Site of his Houses. I have also to announce to you the discovery of the site of the Ancient City of Norumbega."

He went on to construct a thirty-eight-foot-tall stone tower to mark where he claimed to have found the remnants of a Norse fort at the confluence of the Charles and Stony Brook rivers in Weston. Archaeologists, however, have largely disputed Horsford's claims about Norumbega's existence. Author and independent scholar specializing in Viking history, William R. Short, said, "Horsford basically walked from his house, went to the riverbank, found

Statute of Leif Erickson on Boston Commons.
Wikimedia Commons, public domain

rocks, and said, 'Aha! This is a [Viking] house.'"

Still, many residents in the area have enjoyed the Viking theory. In 1887, Boston erected a statue of Leif Eriksson on Commonwealth Avenue. It features a cryptic inscription in ancient rune letters. That same year, telegraph inventor Joseph Barker Stearns built his historic home in Camden, Maine. Norumbega Castle, now the Norumbega Inn, accommodates overnight guests. On top of city hall in Newport, Rhode Island, a weather vane depicting a Viking ship serves as a reminder of Norse pride in the region. In the late 1800s, the city of Bangor, Maine, named its municipal building Norumbega Hall. In 1897, the popular Norumbega Park opened in Auburndale (a village of Newton), Massachusetts. The amusement center closed for good in 1963.

DEVELOPING STORY: THE MYSTERY OF THE RUNE STONES

Several stones with runic letters have been discovered throughout New England. Runes are the twenty-four individual letters of the runic alphabet, which was used by the Vikings and other Germanic peoples from about the second to the fifteenth centuries. While some point to these stones as evidence of the Viking occupation of New England, others remain skeptical of their authenticity. Here are a few of the famous examples:

• In December 1984, a quahogger unearthed an eight-foot-by-five-foot rock off of Pojac Point in North Kingstown, Rhode Island. Two lines of runic letters were inscribed in the boulder. One translation says the stone warns of bears in the area. A local man

Some say that this stone carved with rune letters, resting in Wickford, Rhode Island, is proof that Vikings roamed the region. Photo by Gwen Dane

claimed he carved the characters as a bored teenager in 1964, but others defend that it is the real deal. In 2015, the stone became an official tourist attraction when it was placed in Old Library Park in Wickford, Rhode Island.

• Some believe that three inscribed stones found near Spirit Pond in Phippsburg, Maine, in 1971 were left by Norse explorers. Although dismissed by many as a hoax, the Maine State Museum houses them. Forensic geologist Scott Wolder put forth a whole new theory on the rocks in 2014, saying the stones were created by the Knights Templar who fled Europe for North America after their persecution in 1307, bringing with them the Holy Grail.

• In a shallow well on the grounds of the Tuck Museum in Hampton, New Hampshire, a large rock covered in runic letters is alleged to be the stone that marked the resting place of Thorvald Eriksson, brother of Leif. Several archaeologists have debunked the notion, but tourists still flock to the rock. ■

PART II

ABANDONED AMERICA: COMMUNITIES THAT FLOURISHED AND FADED

THE
HEADLINE

The Anasazi cliff dwellings at Mesa Verde National Park in Colorado are some of the best preserved. The Cliff Palace, shown here, had 150 rooms. Wikimedia Commons, photo by Andreas F. Borchert

LOCATION: **American Southwest**
DATELINE: **1500 BC–AD 1300**

THE VAST ANASAZI CIVILIZATION OF THE SOUTHWEST

Why did the Anasazi abandon their homeland? Did cannibalism play a role?

From as early as 1500 BC to AD 1300, the Anasazi people thrived in an area of the Southwest where Utah, Colorado, Arizona, and New Mexico meet, known as the Four Corners Region. *Anasazi* means "the Ancient Ones," and over hundreds of years these people built a remarkable civilization, despite little rainfall and harsh weather conditions, where temperatures could drop to well below zero during the winter and soar to scorching-hot levels in the summer.

At its peak, the community spread out over forty thousand square miles with roads connecting different settlements. Scientists estimate that the network of thoroughfares covered about four hundred miles. At some points in the desert and canyons, these pathways stretched as wide as thirty feet. This sophisticated civilization developed a rich culture with finely crafted baskets and pottery, crop cultivation, commerce, mathematics, astronomy, calendar-making, and unique architecture. The citizens were scientists, artisans, farmers, hunters, and traders.

But by 1300, the Anasazi had completely vanished from the region. Why did such a vibrant civilization disappear? While some scientists familiar with the region suspect that some type of cataclysmic event compelled the Anasazi to abandon their homeland and move south, others believe that strife among tribes broke the people apart. Some archaeologists think that the Anasazi may have even practiced cannibalism, a ritual that may have contributed to their demise.

How did this great culture develop, and what is the mystery behind the Anasazi's disappearance from the area?

WHAT DO WE KNOW?

Although the Anasazi kept no written history, much has been learned about their culture from archaeological findings, accounts from Spanish explorers, and stories passed down among Native tribes in the region.

Fantastic Farmers

Often called "the Ancestral Puebloans," the Anasazi were an agricultural people known for their skills at cultivating and harvesting crops such as corn, beans, and squash. They even grew popcorn and raised cotton in some areas. With very little rainfall in the area, the Anasazi depended on an agricultural technique called dry-land farming to produce a high yield. Dry-land farming relies on the efficient use of little amounts of moisture trapped in the soil, and the wise selection of crops that will adapt to arid conditions. These farmers constructed irrigation ditches to channel water, dams on mesa tops to capture rainfall and snowmelt, and sometimes located fields near springs. They also learned how to store crops so they would have supplies during seasons when crops produced very little yield.

Constructed of sandstone, wooden beams, and mortar, the cliff dwellings of the Anasazi, or Ancestral Puebloans, were like ancient apartment buildings. Wikimedia Commons, Photo by Andreas F. Borchert

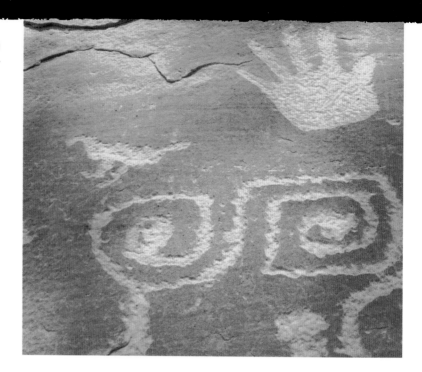

These Anasazi petroglyphs were found in Mesa Verde National Park in Colorado. Wikimedia Commons, photo by James St. John

In addition to eating their corn and squash, the Anasazi also gathered wild fruits, seeds, and nuts. They relied on indigenous trees, shrubs, and other plants to make houses, clothing, medicine, and tools.

Although they hunted for wild game, these ancient people raised turkeys as well. These birds offered a source of meat, eggs, feathers, and bones. Because turkeys eat grasshoppers and other insects, they helped to maintain healthy crops. The Anasazi also kept pet dogs that assisted them in hunting and carrying heavy loads.

Clever Craftsmen

Artifacts have indicated that the earliest Anasazi were skilled basket makers, weaving them with detailed designs. They coated the insides of the baskets with pitch — a dark, tarlike substance that could make baskets watertight. They then advanced to making refined pottery. Hundreds of trench kilns have been found across the Four Corners area. Their pottery often displayed elaborate black-and-white geometric designs. Using yucca plant leaves, the native craftsmen painted the designs with the juice of the beeweed plant, which dried as a black color.

Artisans employed a scraping and coiling technique that showed tremendous precision. The clay cooking pot served as an

The Anasazi created well-crafted pottery often distinguished by black-and-white geometric designs, as shown on these mugs. US Department of the Interior Bureau of Land Management

TIMELINE

1500 BC–AD 200	Anasazi begin to cultivate crops in transition away from nomadic life
200–500	Further cultivation of plants such as corn, beans, and squash. Construction of pit villages. First pottery
500–750	More-elaborate pottery. Construction of great kivas. More-sophisticated weapons. They begin settlements in Mesa Verde region
750–900	Aboveground pueblo adobe structures. Communities move from cliff tops to canyon bottoms
900–1150	Villages expand and road network is built.
1150–1276	Population decline and move to make dwellings carved in cliff sides
1276–1300	Drought

essential item in an Anasazi household, as well as containers for wet storage, dry storage, and serving.

Animal skins became blankets and clothing. Some articles of clothing were sewn of rabbit fur, turkey feathers, and yucca fibers. Sandals were made from yucca leaves.

Archaeologists have found tools and weapons such as stone clubs, daggers, and the atlatl, a device for throwing spears farther and with greater speed. In time, the people learned to fashion bows and arrows for hunting.

Bones and antlers were crafted into needles, combs, gaming pieces, and jewelry. The inhabitants also worked with metals, fashioning copper bells, tools, and weapons.

A Talent for Trade

In 1924, brothers John and Fay Perkins, of Overton, Nevada, stumbled across the ruins of the "Lost City" at a complex of villages near Pueblo Grande. Items discovered here supported the notion that trade and travel were part of the Anasazi livelihood. They developed commerce with other nearby tribes, like the Hohokam, who lived farther south in the hot Arizona desert, and the Mogollon, who were farther south in New Mexico. The Online Nevada Encyclopedia says that much of the pottery found at the Lost City was made in the high mountains of northern Arizona, about eighty miles east. Since the area around the Lost City was fertile for growing crops like corn, researchers think it likely that the high mountain people traded pottery for food grown in the Lost City.

At this one location, scientists also uncovered seashell beads from the Pacific Coast, turquoise from California, and obsidian from Utah, or elsewhere in Nevada—all confirming that the Anasazi had a network of trade. The Anasazi also mined salt caves in the area for trade and food preparation.

A Shift in Housing Revealed a Threat

The early Anasazi lived in pit houses that were partly underground. They laid tree trunks and branches over them to create roofs. But after about AD 750, they constructed pueblos, or aboveground houses made of stone and a heavy clay called adobe. Built with fine masonry, the pueblos often featured multiple-room dwellings resembling apartment buildings. Homes often provided a hearth, fire hole, and places for storage. Built beginning in the ninth cen-

TOP: **Some of the Anasazi homes were only reachable by tall ladders (pictured, above right). They may have pulled the ladders into their homes to protect against invaders, according to some historians.** Courtesy of I Einar Kvaran

ABOVE: **Located in Santa Clara Canyon near Española, New Mexico, the Puye Cliff Dwellings are the ruins of an abandoned pueblo.**
National Archives and Records Administration

tury, Chaco Canyon's Pueblo Bonito in northwestern New Mexico is an amazing example of their architecture. It's estimated that the sprawling structure contained about eight hundred rooms. Parts of the complex were tiered and reached as high as five stories tall.

Originally, Anasazi homes were easily accessible and basically located in the open, but around 1250 there was a notable shift. The people began building cliff housing carved high into the sides of rock cliffs. They were accessible by log ladders. Why did these inhabitants suddenly need such protection and defense? For some scientists, the cliff structures seem to indicate that the Anasazi were under some sort of threat. One story related by a Hopi elder said that the Anasazi may have carved homes high into the cliffs in an attempt to store food and hide away until raiders left. Inside, rooms were dark and quiet, but often made more accommodating by smoothing the walls with mud and painting them.

Conditions were not very hygienic. Archaeologists believe some rooms were used as toilets and trash was sometimes dumped toward the back of the buildings. The people suffered from various diseases, as well as arthritis and bad backs. The rough, cornmeal-heavy diet wore away at teeth, causing cavities and other dental problems.

To conduct religious ceremonies, inhabitants built a round underground room called a *kiva* into each pueblo or cliff house. Inhabitants often gathered together at "great kivas" that served as community centers to perform ceremonies, socialize, and discuss matters of the day.

The Anasazi left behind hundreds of petroglyphs (rock art images), many believed to represent images of celestial phenomena. One image may show a bright star in the sky that scientists believe could reference an AD 1054 supernova. Religious ceremonies involved astronomical movements of the sun, moon, and planets. Ceremonial centers were aligned with solar and lunar cycles, and helped inhabitants to measure the time of year. On the top of Fajada Butte in Chaco Culture National Historical Park, three stone slabs were set up so that sunlight and moonlight pass at the solstices and equinox in such a way as to hit a spiral petroglyph.

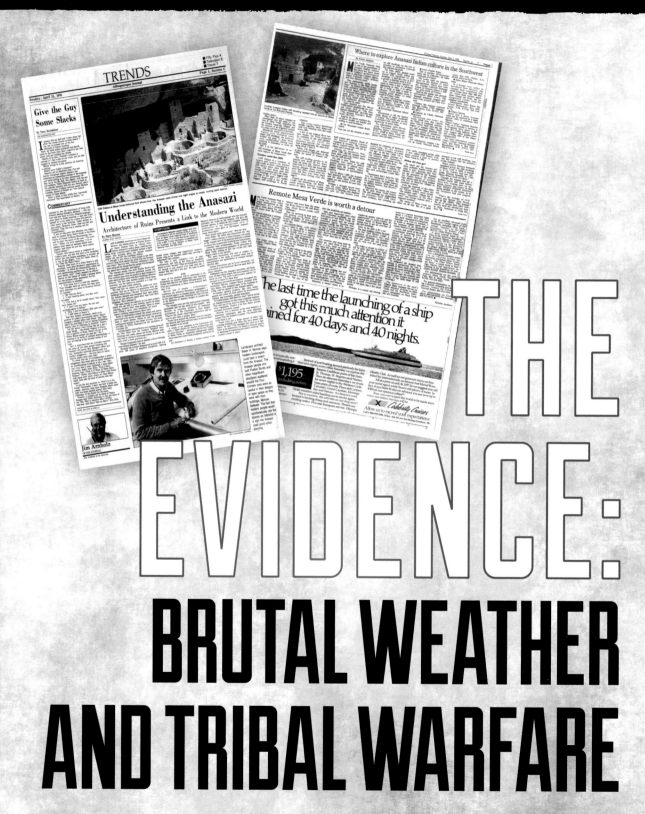

THE EVIDENCE:
BRUTAL WEATHER AND TRIBAL WARFARE

By the late thirteenth century, thousands of Anasazi abandoned their magnificent settlements. They fled their homeland for points south and east, such as at the Rio Grande and the Little Colorado rivers. They didn't just leave behind their homes; they also left their belongings—their pottery, tools, clothing, and other items. In an article in *Smithsonian* magazine, Stephen Lekson, an archaeologist with the University of Colorado, says, "After about AD 1200, something very unpleasant happens. The wheels come off."

Bad Weather

By examining the rings in local trees, scientists found that a massive drought may have crippled the Southwest between 1276 and 1299. Data shows that in some regions there may have been little to no rain for twenty-three years. Some evidence suggests that the Anasazi may have experienced a great famine; after depleting the natural resources in the region, perhaps they needed to move on.

Based on measurements of pollen layers that accumulated in the bottom of lakes and bogs, archaeologists also determined that the weather may have turned colder, making for shorter growing seasons.

Warring Factions

Although early explorers speculated that warring nomadic tribes may have battled with the Anasazi and driven them from their communities, some researchers today doubt that tens of thousands of Anasazi were compelled to leave by force.

Professor Lekson says that remains of executed individuals have been found, including skeletons with skulls bashed in. It appears that Anasazi leaders may have used fear to stay in power, and perhaps killed those who disrupted their rule. They may have even cannibalized them. Lekson conjectures that this type of oppressive leadership created a paranoid and fearful society that could not sustain itself. In time, entire villages attacked each other, and the grand Anasazi society unraveled.

Many historians believe ancient cultures were peaceful, but there is evidence of warfare among tribes. A Hopi elder story says that even the cliff homes could not keep out invaders, and bloody massacres unfolded, with survivors fleeing the area.

Several archaeologists believe that some sort of environmental catastrophe pushed the Anasazi to the brink and that this may have sparked intertribal warfare. The combination of the two factors may have led to the demise of a once-great society.

FRINGE THEORY

Physical anthropologist Christy Turner of Arizona State University published a book titled *Man Corn*, which documents many cases of cannibalism among the Anasazi. Signs of possible cannibalism include cut marks made on bones with stone knives, the breaking of long bones to get at the marrow, the pulverizing of vertebrae, marks indicating that a bone was pounded by rocks, and a sheen left on bones that were boiled. A study published in *Nature* magazine adds further proof of cannibalism: When scientists examined fossilized excrement (coprolite) at one site, they found traces of proteins only found in humans. The researchers believe the only explanation is the consumption of flesh. Traces of human matter were found in cooking pots, as well.

DEVELOPING STORY: AN EXODUS SOUTH AND CLIFF DWELLINGS LEFT BEHIND

What pulled the Anasazi south? No one is exactly sure. Some say they found more-fertile grounds. The settlement at Arroyo Grande, for example, was located above a spring. One theory is that a few Anasazi in the south had developed the Kachina Cult. This was a religion practice centered around spiritual beings called kachinas. The tribes often depict kachinas in the form of elaborately ornamented dolls. Some his-

torians trace the Kachina Cult back to at least 1300, which would correspond with the great migration of the Anasazi. By the middle of the 1300s, Kachina was the focus of the Anasazi spiritual life. Their numbers grew, and they developed huge pueblos, some with twenty-five hundred rooms.

Today, visitors can tour the ruins of the pueblos and cliff dwellings. The Pueblo Indians—the Hopi, Zuñi, Laguna, Acoma, and other Native American tribes—are today's descendants of the Anasazi. They live in communities along the Rio Grande, in New Mexico, and in northern Arizona. Many reside in pueblo-style structures as the Anasazi once did. Some even have kivas. The Pueblo people maintain their traditions in the form of storytelling, language, dance, and crafts, passing them on to their children in an effort to keep this ancient culture alive. ∎

The small, rounded rooms made by Pueblo Indians resemble human swallows' nests dug into a cliff. American Museum of Natural History

THE HEADLINE

LOCATION:
Roanoke Island, North Carolina

DATELINE:
1585–1590

ROANOKE: AMERICA'S FIRST COLONY

Smitten with the dashing Sir Walter Raleigh, Queen Elizabeth I agreed to support his plans to develop a foothold in North America. Public Domain

More than one hundred settlers disappear on Roanoke Island, leaving behind two mysterious clues

When Elizabeth I became queen in 1558, England lagged behind the other major superpowers when it came to expansion and colonization in the New World. France and Spain had already staked out claims to huge portions of land in the Americas, while England had no presence at all.

Sir Walter Raleigh, an adviser to the queen, intended to change that. He persuaded her that it was in the best interest of England to build a colony there.

After an exploratory expedition in 1584, England established a presence on Roanoke Island, in today's Dare County, North Carolina. in the summer of 1585. Although the Englishmen who explored the region made friendly contact with some of the Native American people, they also provoked hostilities among certain tribes. By 1586, nearly all of the settlers had returned home.

Raleigh decided that building a community of families would be the best option for creating a peaceful and thriving colony. In the spring of 1587, more than one hundred men, women, and children set sail for the New World under the leadership of John White.

Sir Walter Raleigh funded expeditions to Roanoke with the goal of setting up the first English colony in North America. Anonymous artist, National Portrait Gallery.

About a month after their arrival at Roanoke Island, the settlers began adjusting to their new home—enjoying the abundance of food, constructing lodging, and making friendly contact with some of the indigenous people while learning to protect themselves from others. To prosper, however, they agreed that they would need more supplies—and more people—from England. In August of 1587, John White sailed back to get Walter Raleigh's help with these requests.

Unfortunately, when White returned, England was at war with Spain, and the New World was not a priority. Despite his continued efforts, White could not get the help that the colony needed. After three years, White finally found a way to return to America on his own, but he discovered that the settlement at Roanoke was completely deserted. There were no signs of violence—just two mysterious clues: the word CROATOAN carved on a post, and the letters CRO scratched into a tree trunk. What had happened to the promising new colony that was to be the beginning of a new English empire in America? The people of Roanoke had completely vanished.

WHAT DO WE KNOW?

T he mystery of Roanoke is part of a bigger world story tied to the Age of Discovery, also called the Age of Exploration. By the end of the 1400s, Europeans had made great strides in ship design and navigation using maps and the newly invented compass.

Several European countries were interested in finding a mythic waterway that passed through North America, and would provide a trade route with Asia.

In the late 1400s, England had made some initial attempts at exploration when John Cabot sailed the eastern coast of Canada in search of a Northwest Passage. His son, Sebastian, traveled the region a few years later, but England did not establish a presence in the New World at this time.

In 1492, Christopher Columbus also went in search of a new route to Asia. The explorer landed in the Bahamas, and his discovery ushered in an era of Spanish domination in Central and South America. In the early 1500s, Spanish conquistadors (Cortes and Pizarro) wiped out the Aztec and Incan civilizations and brought home tons of gold and silver from the Americas. In 1513, Ponce de Léon discovered Florida, and that same year Balboa crossed the Isthmus of Panama and claimed the entire Pacific Ocean for Spain. With riches flooding in from the New World, Spain became the most powerful country in the world.

France, too, staked its claim in the Western Hemisphere. In 1534, Jacques Cartier sailed into the Gulf of St. Lawrence under the French flag, declaring the surrounding territory to be the property of France. In 1541–42, he returned, setting up a short-lived colony, which he called "Charlesbourg-Royal," at the mouth of the rivière du Cap-Rouge. French fishermen and fur traders started thriving enterprises in North America at this time

Tensions Mount between England and Spain

As New France and New Spain grew, England still had no foothold in the New World. In addition, England and Spain were becoming increasingly hostile toward each other in the latter half of the 1500s. Under Queen Elizabeth, England was ruled as a Protestant country, while Spain, under Philip II, was Roman Catholic. Spain maintained that Elizabeth's cousin, the Catholic Mary, Queen of Scots, was the rightful heir to the English throne, and ultimately, Philip wanted to see England under a Catholic monarch. Starting in 1570, Elizabeth kept her cousin Mary imprisoned to protect her control of the throne.

Meanwhile, in the 1580s, Elizabeth supported the Dutch Protestants in the Spanish Netherlands who were rebelling against Philip's Catholic rule. In an attempt to curb Spain's increasing power, Elizabeth encouraged privateers, such as Sir Francis Drake, to launch attacks on Spanish trade vessels and steal their goods as they were sailing back from the Americas.

Mary, Queen of Scots (1542-87) by François Clouet (1510–1572). Wikimedia Commons

Sir Walter Raleigh Pushes for Colonization

Sir Walter Raleigh, as adviser to the Queen, thought that if England established a colony in North America, it could more easily carry out raids on Spanish ships and potentially build wealth, just as the Spanish had done. He also believed that the elusive trade route to the East could be found.

As a dashing and smart young man in his early thirties, Raleigh held great sway over the monarch and cultivated a flirtatious relationship with Her Majesty. Although she was dubbed the Virgin Queen because she never married, or had children, Elizabeth relished the attentions of men and maintained a stable of both domestic and foreign suitors. Smitten with Raleigh, Elizabeth showered him in generosity, giving him estates in Kent and Hampshire—properties that earned him the modern-day equivalent of up to $200,000 a year from fees paid by the vintners who

Francis Drakes portrait in *The World Encompassed*. Wikimedia Commons

A 2017 community theater production on Roanoke Island of *The Lost Colony*, which recreates the saga of the English settlers who vanished. Library of Congress, photo by Carol M. Highsmith

cultivated the grounds. Under the good graces of Elizabeth, he became a wealthy and powerful political force.

The First Exploratory Trip

With a charter from the Queen to occupy, cultivate, and benefit from the lands discovered in North America, Raleigh organized an expedition targeting regions north of Spanish-occupied Florida. Although he did not accompany them, Raleigh sent two ships, captained by Philip Amadas and Arthur Barlowe, to set sail from Plymouth, a port in southwest England, on April 27, 1584. After sailing a well-established route through the Canary Islands and the Caribbean, the two ships reached the Outer Banks of North Carolina on July 4, 1584. The crew rowed their longboat ashore and proclaimed (to no one but themselves) that they were taking

AMIDAS AND BARLOW AT ROANOKE ISLAND, 1584
The Discovery, Exploration & Conquest of America

possession of the land in the name of the Queen. The Englishmen were excited to find that the grounds were alive with deer, rabbits, and wild fowl, plus an abundance of red cedars, pines, cypress, and sassafras trees.

Three days later, the white explorers spied a small boat approaching with three strange Native American men aboard. The men were from the local Secotan tribe. Other tribes in the area were the Weapemeoc, the Chowanoc, and the Croatoan. The English explorers invited the lead scout to join them on one of their ships for a meal. Wanting to create good relations with the local people, the English gave the Indian scout a hat, a shirt, and a few other items.

The following day, Granganimeo, the brother of the local chief, and forty to fifty of his men, came to meet the strange explorers. They sat together and ceremoniously exchanged gifts. The Indians gave the Englishmen deer and bison skins, and they in

Walter Raleigh sent the first expedition to America, headed by Philip Amadas and Arthur Barlowe. The Englishmen established positive relations with the local tribes in Roanoke.
New York Public Library

In 1590, when John White returned to America, he found the Roanoke colony deserted. One of the few clues was the letters CROATAN carved into a post. The Miriam and Ira D. Wallach Division of Art, Prints and Photographs: Picture Collection, The New York Public Library

return gave hatchets, axes, knives, and metal goods. Granganimeo took a special interest in a tin dish. After piercing a hole through it, he hung the dish around his neck, believing it might protect him from his enemies. Soon thereafter, friendly interactions continued, as one of Granganimeo's wives and three or four children visited the Englishmen.

On July 13, on the island called Roanoke, Barlowe and seven shipmates visited Granganimeo's village, which was surrounded by a palisade, a large fence constructed of wooden stakes. The Europeans found a people who were thriving on deer, bear, squirrel, wild fowl, and turkey. They foraged mulberries, huckleberries, persimmons, acorns, and hazelnuts, and cultivated corn, pumpkins, squash, and beans. The Englishmen feasted with the Indians and enjoyed a festive Native ceremony.

After making such a positive initial connection, Amadas and Barlowe agreed to sail back to England in mid-August to share the good news of their discoveries. They made arrangements with Granganimeo to leave behind two of their men; in exchange, they would take two Indians—Manteo and Wanchese—back to England to demonstrate to their countrymen how peaceful these coastal Indians were.

Arriving back in early November 1854, Amadas and Barlowe presented Raleigh with a glowing picture of life on Roanoke Island, although they did warn of some warring local tribes and some distrust of their intentions.

The Second Expedition

Back in Europe, troubles simmered between England and Spain. Philip II continued to plot to overthrow Elizabeth and replace her with a Catholic ruler favorable to Spain. To press the Queen on plans for continuing colonization, Raleigh recruited the help of geographer Richard Hakluyt, who had a great deal of influence over Elizabeth. In 1584, Hakluyt presented an impassioned argument for the political and economic benefits to be gained from such a colony, and the need for England's financial support. While Hakluyt highlighted the abundance of natural resources and untapped riches, including mines of silver and gold, he stressed that communities with large numbers of people would be needed to benefit from all America had to offer.

Raleigh declared that the lands in America should be called "Virginia," after Her Majesty, the Virgin Queen.

Hakluyt also believed that England could set up a stronghold of Protestantism in the New World, convert Indians to Christianity, and counter the spread of Spanish Catholicism. From a strong base in America, the English could also more effectively attack Spain's commercial vessels, which would greatly weaken its empire.

In November of 1584, as the first expedition returned from Roanoke, Raleigh was elected to Parliament. He guided a bill through the House of Commons confirming his plan to colonize America. Raleigh's plans received a boost when Elizabeth knighted him on January 6, 1585. Raleigh declared that the lands in

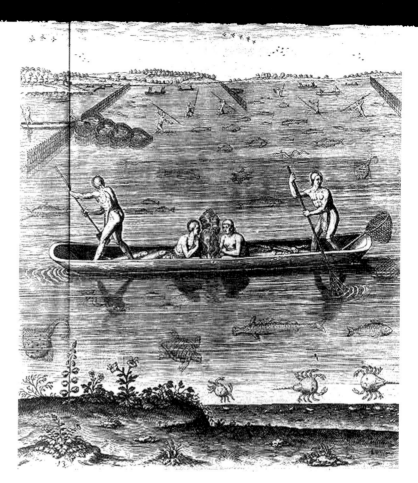

Engraving of John White drawing, depicting Native Americans fishing. Theodor de Bry, Public Domain

America should be called "Virginia," after Her Majesty, the Virgin Queen. In return, Elizabeth gave her beloved Raleigh the title of Lord and Governor of Virginia.

On April 9, 1585, the second expedition prepared to set sail with about six hundred soldiers, sailors, and artisans, led by Sir Richard Grenville. Raleigh again remained behind. The Indians Manteo and Wanchese accompanied the men to serve as liaisons and help to build relations with the local tribes. The artist John White also joined the crew in order to record images and impressions of the new land. At this point, no women or children came along.

At the end of June, Grenville arrived at the Outer Banks, but one of the ships accidentally ran aground. Provisions that were to have lasted a year were now drastically reduced to a twenty-day food supply. The lack of food would have certainly influenced the attitudes of these new arrivals.

Although many Indians envisioned peaceful and profitable trading with the white men, Granganimeo's brother, Chief Wingina, grew suspicious of their intentions. His doubts increased when the Indian Wanchese fled Grenville's expedition and warned that the colonists could not be trusted. Manteo, however, stayed on as a faithful supporter of the English. Furthermore, both Manteo and Wanchese came back to America describing the great stone city of London, overcrowded with people and filth—a vision that many Native Americans did not want duplicated in their ancestral lands.

In early July, relations went smoothly as the Indians welcomed the English with more feasts and ceremonies. White made detailed drawings depicting their dress, dances, and customs. After a visit to one village, however, Grenville believed that a silver cup had been stolen. When it was not returned, he ordered that the village and cornfields be burned to the ground to teach the Natives a lesson. Although Grenville wanted to make it clear who was in charge, news spread quickly to other Indian communities that the English could be violent and dangerous. The episode fueled underlying hostilities toward the white settlers.

Still, with the help of the Indian liaison Manteo, Grenville and Granganimeo came to an arrangement on July 21, 1585, whereby the English would build a community near the north end of Roanoke Island, not far from Wingina's village. The men set to work immediately. Over the course of about a month, they constructed a settlement with a storehouse, magazine, and a chapel, as well as barracks for the men and small single-unit houses for the officers. It's thought that a cooper's shop, a kitchen, a garden, and livestock pens were also part of the new settlement.

By late August, Grenville considered their mission a success and decided to return to England to report the good news. He would sail off with about five hundred of his men, including John White, who would present his detailed drawings of Roanoke to Raleigh. Grenville would leave behind just over one hundred men to look after the new Roanoke settlement under the command of Ralph Lane. A relief expedition with hundreds more was expected to arrive in the autumn.

But when no more ships arrived in the fall, Lane and his men grew restless. As their food supplies dwindled, they became increasingly dependent on the Indians for sustenance. Meanwhile,

Etching of Native Americans cooking fish based on a drawing by John White. Public Domain

the numbers of Secotan who became sick and died multiplied as European illnesses such as smallpox and influenza spread among the tribes. Wingina grew increasingly distrustful of the English. He concluded that the white man had brought death to his people, and they must be eliminated. In the spring of 1586, Chief Wingina prepared to push back against the colonists; the first step was changing his own name from Wingina to Chief Pemisapan ("He who watches closely").

Violence Erupts

Pemisapan plotted to deceive Lane. He convinced him that the Chowanoc and the Mangoak were conspiring to attack Lane and his men, and he should launch a preemptive attack on those tribes. When Lane confronted the Chowanoc chief Menatonon with this story, he told Lane that the Chowanoc had no grievance with him, and that it was Pemisapan who wished to destroy them.

Around this time, Lane also heard rumors of valuable *wassador* (the Indian word for metal) being found further inland. He organized a journey into the hills in search of gold and silver, but after a futile three-week trek, Lane and his team returned to Roanoke empty-handed.

Believing Lane and his group might perish in the woods, Pemisapan was surprised to see them return. He was now determined to cut off their food supply and starve them to death. Lane learned of this treachery and decided upon a preemptive strike. His surprise attack caught Pemisapan off guard. One of Lane's men captured him, and in a heat-of-the-battle frenzy, he cut off Chief Pemisapan's head. While this act put an end to Indian attacks for the time being, it led to great unrest and distrust among the tribes.

By June of 1586, the one-hundred-plus Englishmen at Roanoke were growing increasingly dissatisfied. Unexpectedly, on June 8, twenty-three ships sailed by Sir Francis Drake appeared off the coast. One June 18, Lane and all of his men bid good-bye to Roanoke. Unbeknownst to Lane, however, Grenville was sailing back to Roanoke at this very moment with supplies and two hundred additional settlers. When these new arrivals made it to Roanoke in July and found that Lane and his men were gone, they too decided to reverse direction. They all returned to England. Grenville left just fifteen men with a year's worth of food and a promise that a

Village of the Secotan tribe as depicted in a watercolor by John White in 1585. British Museum, London, Public Domain

A Secotan warrior in North Carolina, as depicted by John White. British Museum, London, Public Domain

John White's drawings of Native American life gave the English a clear picture of what life in America might be like. This watercolor depicts a Secotan warrior ceremony. British Museum, London, Public Domain

TIMELINE

1492	Christopher Columbus claims the West Indies for Spain and opens up the door to Spanish exploration in the New World
1497	John Cabot explores Newfoundland
1519	Cortes begins conquest of Mexico
1532–35	Pizarro's conquest of Peru
1584	Sir Walter Raleigh launches first exploratory expedition to America
1585	Richard Grenville sails six hundred men to America where Roanoke settlement is established.
1587 February 8	Mary, Queen of Scots, executed
July 22	More than one hundred men, women, and children arrive at Roanoke to establish the new colony
November 8	John White arrives back in England.
1588 July 29	Defeat of the Spanish Armada
1590 August 18	White returns to Roanoke to find the settlement completely abandoned.
c. 1593	John White dies
1607	Jamestown, England's first permanent colony, is established

bigger expedition would come soon.

Raleigh was furious at Lane. Instead of helping to establish a new colony, he had abandoned it. But Raleigh was not about to give up his dream. He consulted with John White, who drew elaborate maps of Virginia. Based on White's accumulated knowledge, Raleigh decided that a colony on the Chesapeake Bay would provide a better location than Roanoke. He envisioned a port city named after himself that would rival London and Plymouth. It would become a central harbor for English privateers and a base from which to reach farther into the West, in an effort to find gold, silver, and other valuable minerals and gems. In time, a route to Asia might still be revealed.

The Third Expedition

In the fall of 1586, as plans developed for a new expedition, Raleigh decided against sending just men to the New World. The men alone had become too violent and restless. He concluded that a civilian settlement of men, women, and children would fare far better and create a more peaceful community. Raleigh's pick to lead this colony was John White, who would become the governor of the new city of Raleigh. In January 1587, White and twelve assistants were elevated to the gentry class. Their mission: to create a new small aristocracy in Virginia.

Enchanted by the flora, fauna, and people of America, White bet his life and wealth on the risk. He recruited other families who would join him. To entice English men and women to sign up, Raleigh promised individuals five hundred acres of land and a voice in the new local government.

Gathering together families for the expedition took some time. As White reached out to potential settlers, Mary, Queen of Scots was executed on February 8, 1587. Much of England celebrated—church bells were rung, guns were fired, bonfires were lit, and impromptu feasts erupted in the streets. For many, her death meant that a threat of possible Catholic rule had ended, and increased stability lay ahead. With Mary gone, however, Philip of Spain now viewed himself as the Catholic leader of the English and the Scots, emboldening him to attack England the next year.

After taking care of their affairs and collecting clothing, farming tools, cooking instruments, books, bedding, and other supplies, ninety-two men, seventeen women, and nine children set sail from Plymouth on May 8, 1587, along with about thirty crew. After weeks of treacherous ocean sailing in cramped quarters, with food starting to rot, the settlers made it to St. Croix in the Virgin Islands. Here they found freshwater and enjoyed a feast of roasted turtle meat. Two Catholic travelers among the group, however, deserted the party and disappeared. White feared they

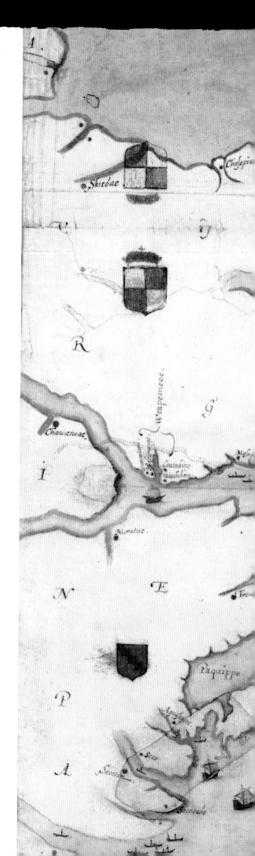

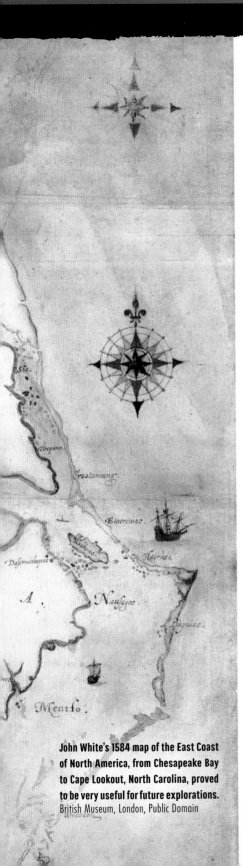

John White's 1584 map of the East Coast of North America, from Chesapeake Bay to Cape Lookout, North Carolina, proved to be very useful for future explorations.
British Museum, London, Public Domain

would reveal their destination to the Spanish. By the third week of July, the new colonists had made it to their final destination and anchored off the coast of the Outer Banks.

On July 22, the newcomers rowed ashore to the Roanoke settlement, eager to connect with the fifteen men who had been left behind the previous summer. To their shock, they found only the bleached bones of the men. They assumed that local Indians had killed them. Although now anxious about possible attack, they were weary from their long trip, and the settlement offered ready-made shelter. The buildings were largely intact, vegetables and melons were growing, and deer roamed about. Instead of forging on to the Chesapeake Bay and launching the new city of Raleigh, they decided to establish themselves at Roanoke first. Once they had settled and adjusted to their new environment, they would move on to Chesapeake.

As they set about repairing the buildings, they admired the fertile and untouched landscape around them. Just days after arriving, colonist George Howe set out alone to catch crab in a nearby creek. He did not know that Secotan warriors were secretly watching him. The Natives launched a surprise attack, riddling his body with arrows and beating him with clubs. When White and the others found Howe's body, they were stunned and worried.

After burying Howe, White and a small group of men went to visit the local Croatoan tribe to try to reestablish friendly relations and determine what had triggered the attack. Manteo, the Indian liaison who had been so helpful in the past, once again smoothed the way. The Croatoan explained to White and the newcomers that many Indians worried that the settlers were destructive—ready to wipe out their crops and take their food. That's why the Secotan warriors had killed Howe—and that's why the fifteen settlers had been murdered.

Manteo made clear that White and the families who had just arrived posed no threat—they only wanted to reestablish peaceful relations. The Croatoan and White came to an amicable agreement. White also asked the Croatoan to send a message to the Secotan Indians, that they wished to forgive all and restart friendly relationships.

After waiting about a week and not hearing any word from the Secotan, White and his men assumed the worst. They decided that their best option would be to launch an attack first. On August 9, men from the Roanoke settlement, along with Manteo, fired upon a group of Indians they thought were Secotan. Shortly into the

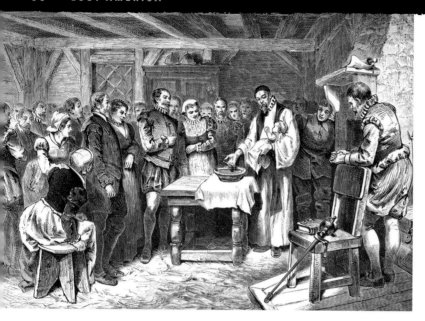

John White's daughter Eleanor gave birth to Virginia Dare, the first English child born in the New World. Image by Henry Howe, Public Domain

battle, they realized they were firing upon friendly Croatoan. Although they patched up relations with the Croatoan, the unfortunate episode only added to the unease.

Deciding it best to focus on their own affairs, the newcomers spent the next few weeks of August in relative harmony, setting up their new homes. They celebrated as White's daughter Eleanor gave birth to the first Christian in Virginia, named Virginia Dare. A few days later, another woman among them gave birth.

The settlers held a christening ceremony for Manteo, bestowing upon him the title of Lord of Roanoke and Dasemunkepeuc. When they moved on to Raleigh, Lord Manteo was to be Raleigh's Indian governor. White assumed that in time the Indians would embrace Christianity as Manteo had, because it was the right path, but many Indians did not want this culture forced on them.

As things progressed for the settlers, they planned to move ahead with the next phase; Roanoke was just a pit stop. Successfully relocating to the Chesapeake Bay, however, would require more supplies and more settlers. The community voted that John White return to England, report the good news (so far) to Raleigh, and then return quickly with needed materials and additional colonists. They also decided that most of the settlers would move inland or nearby, close enough to be protected by the friendly Croatoan people. They would leave a small party behind at Roanoke so White could reconnect when he returned.

If an emergency came up and no one could remain at Roanoke, White instructed them to carve their location in a tree so he could find them. If they had been forced to flee because of hostile Indians and Spanish invaders, the settlers were to make the mark of a cross over the destination to indicate that there had been trouble.

Just over a month after arriving, White left for England on August 27, arriving on November 8. He immediately went to Sir Walter Raleigh to initiate the next stage of colonization.

His timing could not have been worse.

THE EVIDENCE:
A COLONY ABANDONED

Hostilities with Spain were now coming to a head. England was preoccupied, preparing for an invasion. Although Raleigh wanted to proceed with helping Roanoke, a return mission would have to wait.

In March 1588, Grenville prepared to lead a substantial fleet of ships to both attack the Spanish in the Caribbean and to resupply Roanoke, but he was diverted back to Plymouth to battle the Spanish Armada. On April 22, White was able to head off with two smaller ships and fifteen new settlers (seven women, four men, and four children). Unfortunately, an attack by French privateers left White wounded, forcing them to turn back. In July,

Spanish captain Vicente Gonzales discovered Roanoke, but found no settlers and moved on. Had the colonists already abandoned their new outpost at this point?

After the summer of 1588, when Queen Elizabeth led a decisive defeat of the Spanish Armada, White thought he could now get back to America. Raleigh, however, was losing interest in the whole endeavor, and was preoccupied with other matters, including falling in love with one of the Queen's maids, Elizabeth Throckmorton.

Finally, in the summer of 1590, White arranged with some privateers to get transport for himself only. On August 18, White and a few of the sailors traveled up the Roanoke Sound, searching for the English sailors.

John White Finally Returns

When he arrived at the settlement, White was surprised to find that the community had been abandoned. He was relieved to see that there were no signs of harm or violence. If there had been an Indian or Spanish attack, famine, or disease, surely he would have found evidence of bodies or graves, but there was none.

The houses had been taken down and a palisade-like structure had been erected, creating a type of fort. Some remnants found scattered about included a couple chunks of lead, some small weapons and ammunition, and some other gear. White uncovered a few chests of personal belongings that he had hidden in a trench; many of the items appeared to have been destroyed, ransacked, or ruined by weather.

The biggest clues to the missing colonists were two carvings that White found: CRO was etched into a tree in "fair Roman letters," and the word CROATOAN was carved on one side of the palisade's main gatepost. The distress signal of a cross was not there. Croatoan Island was the home of Manteo's friendly tribe. Called Hatteras Island today, it was fifty miles south of Roanoke. White assumed that the colony members had moved there.

Captain Abraham Cocke, who had delivered White to Roanoke, agreed that they would search Croatoan Island that next day. Overnight, a ferocious storm swept in, driving one ship

aground. Captain Cocke deemed it too un-
safe to stay, so they sailed off to the Caribbean with a promise to return as soon as possible. The return trip to Roanoke was not to
be, however. Cocke and his men joined other privateers in the Caribbean, and as foul
weather persisted, they decided to sail home
to England, with John White in tow.

Arriving back in England in October of
1590, White saw that his prospects of seeing
his daughter, granddaughter, and all the other
settlers were growing dim. Sir Walter Raleigh's
interest continued to wane, especially as his
affair with Elizabeth ("Bess") Throckmorton
blossomed. In July 1591, Bess became pregnant, and in November, the couple secretly
married. For months, he managed to keep
the marriage a secret from the Queen, who he
knew would lash out against them both.

When Throckmorton gave birth to their
child in the spring of 1592, their secret was
revealed. Queen Elizabeth was so outraged that she imprisoned
them both in the Tower of London. Raleigh bought his release
with profits from a privateering voyage in which he had invested.
Bess gained her freedom in December, but while imprisoned, their
son had died of the plague. When John White passed away the
following year, plans for a search to find the Roanoke colonists
died as well.

Gradually, Raleigh rebuilt his relationship with Queen Elizabeth. He also became obsessed with finding the fabled lost city
of El Dorado and its supposed vast riches of gold. In 1595, he per-
sonally led an expedition some four hundred miles into Guiana,
a region in northeastern South America, but he did not uncover
any gold or a lost city. In 1602 and 1603, Raleigh sent two small
exploratory expeditions back to Virginia, but they failed to even
get close to Roanoke. After Queen Elizabeth died on March 24,
1603, James I (the king of Scotland, and the son of Mary, Queen
of Scots), took over the monarchy in England. He never liked Raleigh so Sir Walter's prestige and influence swiftly declined.

When disease and starvation ravaged his tribe, Chief Wingina turned against the English settlers. Drawing by John White, National Park Service, Public Domain

THE DREAM AND THE SEARCH CONTINUE

The dream of creating an English settlement in North America did not die with Roanoke. On May 14, 1607, a group of about one hundred members of the Virginia Company founded the first permanent colony on the banks of the James River, sixty miles from the mouth of Chesapeake Bay. These colonists were instructed to be on the lookout for any Roanoke survivors. When they first arrived, they spotted a blond, light-skinned young boy along the river, but he did not appear again. They also sent some scouts to the south and heard reports from Indians about clothed men and women, indicating they were European. Still, they did not find any Roanoke survivors.

Jamestown almost collapsed in its first two years due to famine, disease, and conflict with the local Indians. But when new supplies and a new group of settlers arrived in 1610, the community began to thrive, and they soon built a prosperous tobacco trade. As a member of the governing council of Jamestown, Virginia, John Smith led two voyages on the Chesapeake Bay. According to Smith's exploration and interactions with Native tribes, the survivors of Roanoke dispersed and settled in Panawicke, Ocanahonan and Pakerakanick in the North Carolina interior.

With the failure of Roanoke, Jamestown in Virginia became England's first permanent colony, where many practiced productive trades, such as glassblowing. National Park Service

DEVELOPING STORY: ARCHAEOLOGICAL CLUES

Evidence that the Roanoke settlers first moved to modern-day Hatteras Island has been found in archaeological digs there. Some scientists believe that a ring engraved with a prancing lion or horse, excavated from the Cape Creek site on Hatteras, may have belonged to a prominent member of the Roanoke colony. Discoveries on Hatteras jibed with the writings of the explorer John Lawson, who visited the area in 1700. He recounted meeting a group of Hatteras Indians who said that some of their ancestors were white people. He described these Indians as having distinctive gray eyes—common to their people, but to none other.

In an article in *National Geographic* in 2015, Mark Horton—an archaeologist at Britain's Bristol University, who headed the Hatteras excavation—said, "The evidence is that they assimilated with the Native Americans."

At a site in Merry Hill, North Carolina, archaeologists uncovered European ceramics from the Roanoke time period. Other archaeologists at Albemarle Sound near Edenton, North Carolina, have unearthed pottery that they believed belonged to the settlers of Roanoke. "It's not unreasonable to think that some of the lost colonists could have been adopted by Native Americans, who often did accept [outsiders]," says Dr. Stanley Knick, director of the Native American Resource Center at the University of North Carolina–Pembroke.

Some tales passed along by local tribes indicate that the Roanoke survivors lived with the Indians for twenty years, but were then exterminated by the Powhatan, a powerful people who were expanding their rule over Tidewater, Virginia. Still, stories persisted that at least some Roanoke colonists lived on, and efforts continued to find them. Searchers felt that any survivors might have intimate knowledge of the region, and might even know of the rumored treasures that Raleigh had dreamed of.

Some believe descendants of the colonists now reside in Robeson County, North Carolina. The Lumbee people, a tribe with more than forty thousand members, have lived there for centuries. Proponents of the Lost Colony-Lumbee connection point to how Anglicized the tribe became. The Lumbee people have long spoken English and followed Protestant religious traditions.

FRINGE THEORY: ALIEN ABDUCTION

The prevailing theory behind Roanoke is that the people assimilated into local tribes. Because no trace of the actual settlers has ever been found, a few fringe theorists believe that the colony disappeared in one fell swoop in a mass alien abduction. How else could they have vanished and not left any evidence of their existence?

LASTING IMPACT: ROANOKE STILL CAPTURES THE IMAGINATION

Thousands of visitors come to Roanoke Island every year, both for the natural beauty that originally won over the early settlers, and because of the enduring mystery of the lost colony.

The spell that Roanoke has cast can be seen in the many rumors surrounding its legend. It is said that on his deathbed, Edgar Allan Poe in his delirium whispered: "Croatoan." The word is reportedly scribbled in a journal of Amelia Earhart's. It was supposedly seen carved into the bedpost where author Ambrose Bierce slept before his vanishing in 1913. Stagecoach robber Black Bart is said to have scratched it on the wall of his cell before he was released from prison in 1888—never to be seen again. According to rumor, the last page of the logbook for the *Carroll A. Deering* ship was the word "Croatoan." In 1921, the ship ran aground on Cape Hatteras, with no one aboard.

The impact of Roanoke can be seen in modern culture today. TV shows such as *Supernatural*, *Sleepy Hollow*, and *Mind Hunters* have all referenced it. In 2016, the cable network FX debuted the sixth season of its anthology drama series, *American Horror Story: Roanoke*. ■

The television show *American Horror Story: Roanoke* was inspired by the mystery surrounding the famous lost colony. FX Network/ Photofest

THE HEADLINE

LOCATION:
Bodie, California

DATELINE:
1876–1945

BODIE: ONE OF AMERICA'S MOST FAMOUS GHOST TOWNS

How a thriving mining community went from boom to bust

L ocated seventy-five miles southeast of Lake Tahoe, Bodie is one of the most famous ghost towns in America. The old gold-mining town that thrived toward the end of the nineteenth century now looks like an abandoned movie set for an old Western. While several "ghost towns" exist around the United States, Bodie is remarkable in that it is largely unreconstructed—not rebuilt to re-create the Old West, but made up of more than one hundred original buildings that remain mostly intact, including the church, post office, saloon, county barn, jail, general store, and barbershop. The artifacts that were left behind remain untouched—covered in dust, but as they were when the town was inhabited. Pool tables are still set to be played and kitchen drawers remain filled with cutlery. Mattresses lie on the beds. Cans of food line the windowsills.

Bodie hit its peak as a bustling mining town in the late 1880s, but as the gold supplies began to decline, settlers began to seek their fortunes elsewhere. Gradually, the population dwindled. After the turn of the century, mines started to close and local businesses followed suit. *The Saturday Evening Post* labeled Bodie a ghost town in 1915. The town struggled on for a few more decades, but after 1942, with the local post office closing, the community was almost completely gone.

Bodie is the biggest unrestored ghost town in the West. At its height in the late 1800s, it grew to a population of ten thousand. Wikimedia Commons, (photo by MasJess)

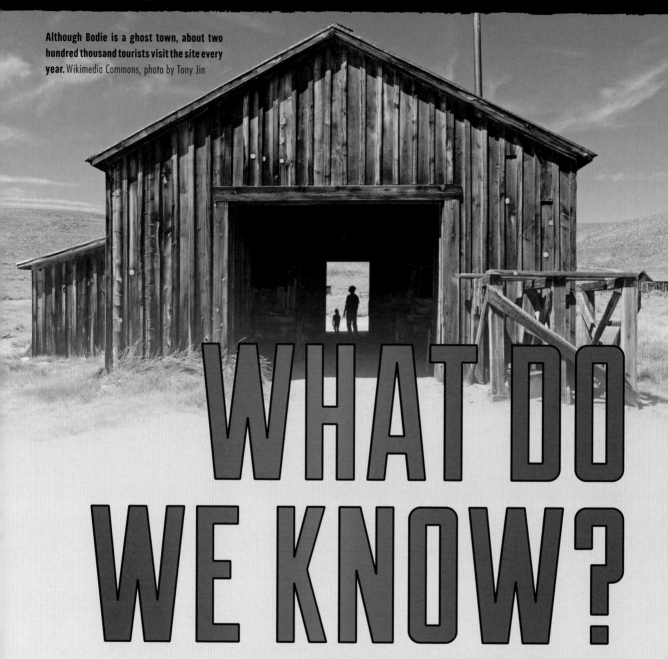

Although Bodie is a ghost town, about two hundred thousand tourists visit the site every year. Wikimedia Commons, photo by Tony Jin

WHAT DO WE KNOW?

Bodie's road to fame starts back in 1848 when W. S. Bodey (possibly William or Waterman Bodey) of Poughkeepsie, New York, left his wife and two kids behind when he heard about prospectors finding gold in California. He quickly headed west in search of riches and adventures, landing in San Francisco in 1849. He pursued his dream for a decade, until he finally found gold in eastern California, north of Lake Mono, near an area now called Bodie Bluff. His good luck wasn't to last long, however. As the winter snows came in November of 1859, Bodey and his Native American companion left their camp to get supplies. The two got caught in a blizzard, and Bodey froze to death.

Although Bodey didn't get to see the booming town that would be named in his honor, he opened up the gateway to a gold rush in the region. According to legend, the spelling of the town changed to Bodie because of a sign painter's mistake.

The community began slowly with the establishment of the Bunker Hill Mine and a mill in 1861. Only about twenty miners resided there for about a decade and a half.

The remnants of the Dechambeau Hotel and the Independent Order of Odd Fellows Building stand at the south end of Main Street in Bodie, California. Wikimedia Commons, photo by Tony Jin

THE BAD MAN FROM BODIE

During its heyday, Bodie gained a nationwide reputation for brawls, robberies, murders, gunfights, and general criminal behavior and chaos. Its reputation equaled Dodge City, Tombstone, and Deadwood. Although the city didn't have one single famous outlaw like Wild Bill Hickok, stories of a legendary frontier villain spread about the "The Bad Man from Bodie." He was a character of folklore like Paul Bunyan that rose from all the tales of general anarchic behavior. *The San Francisco Bulletin* maybe said it best: "The Bad Man from Bodie is a sort of a generic term for all the bad men in the State."

In the 1880s, the "Bad Man from Bodie" gained national recognition when the writer who coined the phrase published a tale in *Sacramento's Daily Bee*. He wrote of one incident when the infamous Bad Man jumped up on a pool table in a saloon and roared to the crowd: "Here I am again—a mile wide and all wool. I weigh a ton, and when I walk, the earth shakes. Give me room and I'll whip an army. I was born in a powder house and raised in a gun factory. I'm bad from the bottom up and clear grit plumb through. I'm chief of Murdertown, and I'm dry! Whose treat is it? Don't all speak at once, for I'll turn loose and scatter death and destruction Hell bent for election. Your treat, is it? Well, come up everybody. Pass the old rat poison."

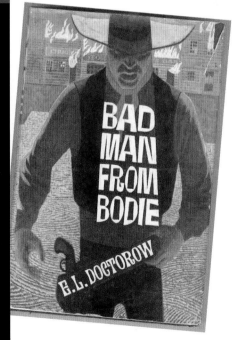

The California Gold Rush began on January 24, 1848, when gold was found by James W. Marshall at Sutter's Mill in Coloma, California. This photo shows miners that year, eleven miles away in El Dorado. Library of Congress

THE EVIDENCE:
BOOM TIMES IN BODIE

While the mine and mill saw several different owners during this time, fortunes rose considerably in 1875 after a mine collapse. The accident revealed a vast vein of gold ore. In 1877, Standard Mining Company took over the Bunker Hill Mine. Uncovering a wealth of gold—and silver—opened up the floodgates. People streamed in, and what was just a town of a couple dozen quickly swelled to five thousand. Many worked for the mine, but others struck out on their own with shovels, picks, and pans, sifting the streams for nuggets and flakes that washed into the running waters.

TIMELINE

1849 — W. S. Bodey lands in San Francisco, seeking his fortune as a prospector

1859 — Bodey finds gold in the area, but dies a short time later in a snowstorm

1861 — Bunker Hill Mine established

1862 — The first spelling of "Bodie" noted

1877 — Standard Mining Company begins operation

1878 — Daily mail service starts in Bodie. Population hits five thousand

1880 — An estimated thirty mines are operating. Population hits ten thousand

1881 — The railway begins delivering lumber

1882 — Population drops to three thousand

1887 — The Bodie Mine and Standard Mine merge

1892 — Devastating fire sweeps through Bodie

1892 — Electricity arrives

1913 — Standard Mining closes

1917 — Town suffers another destructive fire

1940s — Residents number only six by the end of the decade, and Bodie is soon officially a ghost town

1962 — Bodie becomes a State Historic Park

Many of the interiors in Bodie buildings have been left untouched, preserved in a state of "arrested decay." Library of Congress, photo by Carol M. Highsmith

Over the next twenty-five years, Standard Mining raked in about $15 million. As the money flowed in, so did all manner of businesses, gamblers, and prostitutes. By 1880, an estimated thirty gold mines had opened. The community now ballooned to

The Independent Order of Odd Fellows Hall (right) still stands next to the old DeChambeau Hotel. Library of Congress, photo by Carol M. Highsmith

almost ten thousand (mostly young males), with two thousand buildings constructed to house the huge influx. Bodie had become large enough to support three newspapers, a couple banks, and a few churches. There was a racetrack on one end of Main Street and a "Chinatown" on the other. Three breweries and a whopping sixty-five saloons kept hardworking miners happy in their off-hours, along with numerous brothels, gambling dens, and opium parlors.

Good-bye, God! We're moving to Bodie.

Even though Bodie became prosperous, life here was challenging. Street fights, holdups, and killings were commonplace. The Western village gained a reputation for lawlessness. It was

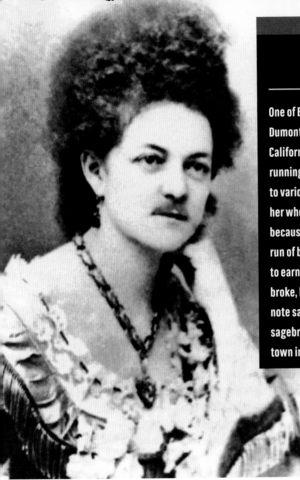

MADAME MUSTACHE

One of Bodie's most famous inhabitants was Madame Mustache. Originally Eleanor Dumont, the educated and fashionably dressed woman arrived in Nevada City, California, in 1854 at about the age of twenty. She started a career dealing cards and running gambling establishments. Popular among the gold miners, Dumont traveled to various booming mining towns and camps, bringing her gambling business with her wherever she traveled. In later years, she got the nickname Madame Mustache because of the prominent dark hair on her upper lip. After a failed marriage and a run of bad luck, she settled in Bodie in May of 1878. Just over a year later, she tried to earn an income offering betting games in the Magnolia Saloon. She soon went broke, however. Despondent over her failure, Madame Mustache wrote a suicide note saying she was tired of life, then followed the road from town, stepped into the sagebrush, and swallowed a lethal dose of morphine. She died a mile or so south of town in September 1879.

Madame Mustache was one of Bodie's most infamous residents. She was a popular card dealer, well-known for the hair on her upper lip. Public Domain

so rough and rowdy that Reverend F. M. Warrington labeled it "a sea of sin, lashed by the tempests of lust and passion." Reportedly when one family was heading to settle there, their little girl kneeled down and said in her prayers, "Good-bye, God! We're moving to Bodie." In addition to the rampant rowdiness, winters could be brutal, and mine deaths from collapses and gunpowder explosions were frequent.

Because the town lacked harvestable trees for construction, building a railroad proved to be a crucial step toward keeping Bodie alive. Enormous amounts of timber were also needed to fuel mining steam engines, support mining tunnels, heat buildings, and cook. Beginning late in 1881, the Bodie Railway & Lumber Company delivered cordwood and timbers from south of Mono Lake to Bodie's mines and mills.

DEVELOPING STORY:
THE MINES RUN DRY

E ven as the trains came in and people rushed to open mines to take advantage of the boom times, it was apparent within a few years that there was not enough gold and silver in the region to support them all. Some stayed open only for a year or two, but then closed. In some cases, miners would dig deeper and deeper only to find low-quality ore that wasn't really worth the cost of excavating. Only a handful of mining companies could survive on the available ore supply. So just as quickly as they came, people started to leave, heading to other, more-prosperous mining towns. By 1882, the population had plummeted to three thousand. In February of 1883, the *Evening Miner* wrote: "A quiet town is Bodie today." In 1887, the two major mines—the Bodie and the Standard—merged. In 1892, a devastating fire helped to hasten the town's demise.

In an effort to reduce costs in 1892, the Standard Company arranged for long-distance electric transmission lines to bring in cheap power from a hydroelectric plant in the Sierra foothills, about twelve miles away. A newfound process called *cyanidation*, which allowed miners to extract precious metals from rock, helped miners to earn more money for their efforts, but even the new technology couldn't pump new life into Bodie.

By the turn of the century, a handful of mines remained in Bodie, and about eight hundred people worked there. But even these soon began to close, and the original Standard Company shuttered its operation in 1913.

The town wasn't entirely dead yet. A couple of mining operations and prospectors came through, trying to dig some final scraps of wealth out of the land. But the railway stopped running in 1917, and another big fire added to Bodie's mounting troubles. In 1929, the Red Mill Mine reopened, but the company couldn't turn a profit, and it closed two years later. Through the 1930s, the Depression and a fire reduced the town to only a few people, who continued to live there until after World War II.

By then the town was down to just six residents—five of whom faced untimely deaths. One man murdered his wife. In retribution for the wife's murder, three other male Bodie residents then killed the husband. Those three keeled over soon after from mysterious diseases. By the end of the 1940s, it was official: Bodie was a ghost town.

Panning was a common method for finding gold in the streams of California during the Gold Rush. The technique involves filling a pan with water, soil, and gravel. The miner then swirls the pan, sending lighter material into the water, while heavier stones and ore such as gold stay in the pan. Wikimedia Commons

THE GOLD RUSH

On January 24, 1848, James W. Marshall found gold at Sutter's Mill in Coloma, California, near Sacramento. Word spread fast, and the Gold Rush was on. The promise of finding riches in the ground and waters of California attracted three hundred thousand people. Everyone had gold fever. In San Francisco, many employees in shipyards, newspapers, and elsewhere dropped everything, abandoned their posts, and headed to the hills to strike it rich. It's estimated that three-quarters of the city's male population vacated. The huge influx of hopeful miners who arrived in 1849 were often referred to as the 49ers (a name the San Francisco football team retains today). Bustling gold-mining towns popped up all over California. The gold that was easy to find on the surface mostly disappeared after 1850. The mining process swiftly became industrialized, driving most of the independent miners to quit their pursuits, many becoming wage laborers in the mines.

LASTING IMPACT :
A THRIVING GHOST TOWN

For a couple of decades, Bodie was largely forgotten, except for some curiosity seekers who would pass through, interested in the remnants of the Old West. In 1962, California made Bodie a State Historic Park. The former post office and saloon were transformed into a museum and visitor center. Still, getting to Bodie isn't easy. Tourists have to travel a dusty, bumpy, thirteen-mile road off of Highway 395 in Mono County to visit the remaining buildings, most of which are kept in a state of "arrested decay"—meaning the structures are protected, but not restored. When many of the inhabitants left after the fire of 1932, they packed their wagons with all of the essentials they could fit, but still left many items behind. Now, the only living people in

NOT QUITE A GHOST TOWN: BUFORD, WYOMING—POPULATION: 1

Buford, Wyoming, has gained a certain amount of fame by being the smallest town in America. Don Sammons claimed the title of sole resident starting in 2008. When Sammons moved on to live with his son, the town was bought at auction for $900,000 by a Vietnamese investor named Pham Dinh Nguyen. Nguyen owns the town's only business—the Buford Trading Post. The store sells all the usual convenience-store supplies, plus bags of Vietnamese coffee that Nguyen imports and sells under his brand name "PhinDeli." But Nguyen isn't the current sole resident. The title—as of this writing—goes to the store manager, Brandon Hoover. Hoover said he wanted to escape the rat race, and Buford has certainly given him that opportunity.

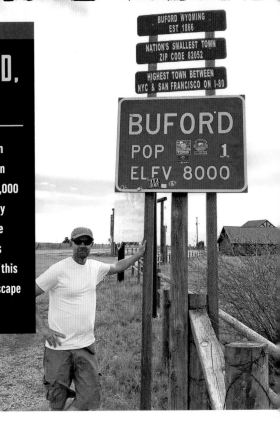

Photo by Rich Lee, used with permission

Bodie are the rangers and visitors who come during the day.

Rangers warn tourists about the Bodie Curse— how misfortune will befall anyone who takes an item from the town, and that bad luck will doggedly follow the individual until the item is returned. Some think the curse is just a ploy to keep people from pilfering items, but a few stories have people believing otherwise. Some visitors have eagerly returned items with letters of apology and descriptions of the unfortunate events that followed. They all hoped to set things straight in their lives by returning the objects they had taken.

After two girls snatched some money from a bed where tourists would toss change, their family suffered money troubles. In the 1990s, one fellow pocketed a few small items. The next year he was in a car wreck, lost his job, and suffered from a chronic illness. On the TV series *Beyond Bizarre* (2000), a German man shared the story of his uncle who had taken a small bottle. Two days later, he crashed his car on the Autobahn. After his son took the bottle to school to show classmates, he had a bicycle accident on the way home. In the 2000s, two teenage girls took rocks from Bodie to make necklaces. They developed rashes where the stones touched their skin. One sprained her ankle. After both lived through an earthquake, they returned the rocks.

Calico is another California ghost town that was once a bustling mining community. Wikimedia Commons, photo by C. G. P. Grey

RIGHT: **Shaniko, Oregon, was once the "Wool Capital of the World." When railroad service was discontinued in 1942, however, the town died out. The town has a collector car museum at its center.** Photos by Greg McCorkle

OTHER GOLD RUSH GHOST TOWNS

Many mining towns sprang up during the mid-nineteenth-century Gold Rush. They filled up with miners seeking their fortunes, but when precious metals were depleted, people left just as quickly as they had come, often leaving behind towns that are still standing to one degree or another today. Here are a few other Gold Rush ghost towns:

Kennecott, Alaska

Pearce, Arizona

Ruby, Arizona

Calico, California

Coloma, California

Animas Fork, Colorado

St. Elmo, Colorado

Bannack, Montana

Rhyolite, Nevada

Virginia City, Nevada

Shaniko, Oregon

South Pass City, Wyoming

FRINGE THEORY:
REAL GHOSTS NOW RESIDE IN BODIE

Bodie is a ghost town haunted by ghosts. A gravestone in the Bodie cemetery marks the resting place of a three-year-old girl. Some visitors have reported hearing the child giggling and playing. Ted Foxx, Alamy Stock Photo

The term *ghost town* is typically used to describe a place where the buildings still stand but where no one lives anymore. It doesn't mean that ghosts actually inhabit the place. Bodie's case is different, however, as there have been numerous sightings of apparitions in the years since the town was abandoned.

Bodie was once home to a sizable Chinese population, many of whom worked on the railroad. One spirit who has made several appearances is a heavyset Chinese maid. She resides at the corner of Park and Green streets, where the J. S. Cain House stands. One story holds that the maid took her life after having an affair with Mr. Cain. The specter reportedly loves children but is not so keen on adults. A ranger's wife who stayed in a lower bedroom one night wrote in 1990: "I was lying in bed. . . . I felt a pressure on me, as though someone was on top of me. I began fighting. I fought so hard I ended up on the floor. It really frightened me. Another ranger who had lived there, Gary Walters, had the same experience, in the same room, except that he also saw the door open and felt a presence and a kind of suffocation."

At the Bodie Cemetery, a striking white marble angel marks the grave of a three-year-old girl named Evelyn Myers, who reportedly died in a horrific manner in 1897. Supposedly, the young girl befriended a local miner, whom she would follow to work many mornings. One stormy day, he told her to go home. She decided instead to sneak up on the miner as he worked. As the laborer swung his pick—unaware that the girl was behind him—he accidentally cracked her in the head with it. One visitor paid a visit to the graveyard with his little girl. When they came to the "Angel of Bodie" grave, the father's little girl started giggling and playing with what seemed to be an unseen spirit.

On some occasions, rangers swear they have smelled delicious aromas of Italian cooking coming from the kitchen of the Mendocini House. It's assumed the Mendocinis kept a happy home, as people have reported hearing children laughing and party sounds.

At the Gregory House, some have been spooked by the sight of an old woman in a rocking chair, knitting an afghan. At times, the chair is empty, rocking on its own. On occasion at the Dechambeau House, a woman has been spotted peering out of a second-story window. ■

THE HEADLINE

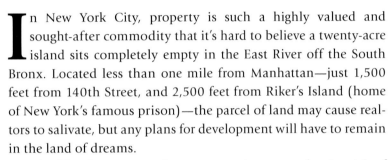

LOCATION:
North Brother Island,
New York City

DATELINE:
1600s – 1963

NORTH BROTHER ISLAND: NEW YORK CENTER FOR QUARANTINING THE SICK

Site of the worst loss of life in New York City before 9/11

In New York City, property is such a highly valued and sought-after commodity that it's hard to believe a twenty-acre island sits completely empty in the East River off the South Bronx. Located less than one mile from Manhattan—just 1,500 feet from 140th Street, and 2,500 feet from Riker's Island (home of New York's famous prison)—the parcel of land may cause realtors to salivate, but any plans for development will have to remain in the land of dreams.

The island once served as a quarantine center for the sick of New York, and then lived on for a short time as a home for veterans and a drug treatment center for teens. Starting in the 1960s, however, the island was completely abandoned. Since the New York City Parks Department gained control of the island in 2011, its only residents have been birds. With its crumbling dormitories and hospital buildings, North Brother exudes a "haunting melancholy, with distant views of the Manhattan skyline," wrote Lisa Foderaro of the *New York Times*.

The Riverside Hospital tended to patients with highly contagious diseases, such as Typhoid Mary. Wikinedia Commons, photo by Reivax

WHAT DO WE KNOW?

In the early 1600s, a Dutch captain, Adriaen Block, claimed North Brother and adjacent South Brother Island for the Dutch West India Company. They were originally named *De Gesellen* (which translates to "the companions"). According to the book *The Other Islands of New York*, by Sharon Seitz and Stuart Miller, settlers did not favor these landmasses because of the treacherous currents swirling around them.

ABOVE: **A portion of a three-story spiral staircase and left, a bedroom at the nurses' residence on North Brother Island.** Photos by Jonathan Haeber, reprinted with permission

A print advertisement dating back to 1791, however, announced that the "Two Brothers Islands" were for sale, touting that their location on the river made them ideal "for a pilot or a house of entertainment." The ad described North Brother Island as having a "dwelling house, barn, orchard, and a variety of fruit trees, with a quantity of standing fire wood and timber." In 1869, a lighthouse was built there to help guide ships through the dangerous waters in the East River. In 1881, North Brother, which was part of Westchester, was transferred to New York City. In August of that year, the *New York Tribune* wrote that the island was occupied "only by a lighthouse-keeper and his assistant, and by a woman who entertains occasional picnic parties."

TIMELINE

Early 1600s	Dutch West India Company claims North and South Brother Islands
1869	North Brother Island becomes part of New York City
1885	Riverside Hospital constructed
1892	Seventy individuals with typhoid are quarantined at North Brother
1904, June 15	The *General Slocum* steamship catches fire, killing 1,021 on board—the largest New York City disaster until September 11, 2001
1907	Mary Mallon (aka, Typhoid Mary) held at the island
1915	Typhoid Mary breaks terms of her release and is again sent to North Brother, where she lives until her death in 1938
1950s–1963	Island serves as teen drug treatment center
2001	New York City Parks Department acquires the land, and North Brother becomes a bird sanctuary

THE EVIDENCE:
AN ISLAND FOR SICK SOULS

In the 1800s, America beckoned hordes of struggling and poor immigrants with its promise of freedom and opportunity. While the immigrants brought many talents to the country, they also brought diseases, including yellow fever, cholera, typhoid, influenza, and tuberculosis.

Because millions of these immigrants entered the country through a processing station on Ellis Island, New York City became a hotbed for illness. In the 1850s, when smallpox endangered lives of city dwellers, New York constructed a smallpox hospital on Roosevelt Island in the East River to keep the sick from the healthy.

As other dangerous ailments threatened public health, the city built Riverside Hospital on North Brother Island in 1885. As a *lazaretto*, this health facility isolated people with infectious diseases. The city shuttled hundreds with deadly and highly communicable illnesses off to North Brother Island, where they could be treated and recuperate without threatening the rest of the city population. Some have said that these quarantine outposts were more effective as public relations tools than anything else. Although they may have helped to a degree in stopping the spread of disease, they were most effective at soothing the fears of the public, who demanded that some action be taken.

North Brother Island housed patients with tuberculosis, an infectious disease that affects the lungs. By the start of the 1800s, the disease had killed one out of seven people who had ever lived. North Brother Island also took in individuals with typhus, an infectious disease transmitted

The nurses' residence at the quarantine hospital on North Brother Island, where patients with communicable diseases were kept isolated from New York City. Photo by Jonathan Haeber, reprinted with permission

by fleas, mites, lice, and ticks. The illness can cause abdominal pain, backache, fever, headache, nausea, diarrhea, delirium, and death.

In 1892, a boat loaded with Russian immigrants docked at Ellis Island. Although the passengers were filthy from their long journey and their bodies infested with lice, officials admitted them. The newcomers headed to family homes and boardinghouses in the Lower East Side of Manhattan. As typhus broke out in this community, many of these immigrants were later rounded up and quarantined. About seventy were shipped to North Brother Island with typhoid fever.

LANDIS & RUPPERT

NORTH BROTHER'S LITTLE SIBLING

South Brother remained the less-popular, smaller sibling to North Brother, measuring just seven acres compared to North Brother's twenty. While it's thought to have served as a Union base during the Civil War, the city used it as a dump through much of the 1800s. Around 1894, Jacob Ruppert Jr. bought the Island. Ruppert gained a fortune as the owner of the Jacob Ruppert Brewery, located between 90th and 94th streets on Second and Third avenues in Manhattan. The brewery pumped out one of the best-selling beers in America at the time. Ruppert also went on to own the New York Yankees starting in 1915. Plus, he has the special claim to fame of acquiring Babe Ruth's contract from the Boston Red Sox. Ruppert built a yacht house on South Brother, and a few informal baseball games were held on a lot on the property. After Ruppert died in 1939, ownership of South Brother changed hands several times, until the investment company Hampton Scows bought it for just $10. It turned out to be a very profitable investment. In 2007, the National Oceanic and Atmospheric Administration's Coastal and Estuarine Land Conservation Program purchased the island for $2 million using federal grant money, and then donated it to the city's Parks Department as a wildlife sanctuary. Today South Brother is a dense forest, covered in vines and flocks of wild birds.

Brewery magnate and New York Yankees owner Jacob Ruppert (right), bought South Brother Island in 1894. Library of Congress

Babe Ruth may have played informal baseball games on South Brother Island. The property was owned by Jacob Ruppert, the same man who acquired Ruth from the Boston Red Sox.
Library of Congress

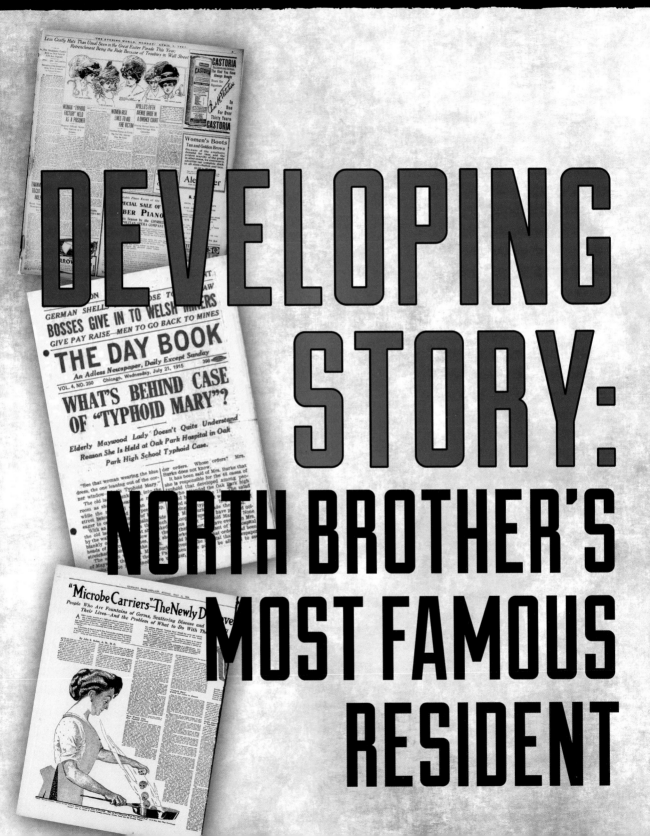

DEVELOPING STORY: NORTH BROTHER'S MOST FAMOUS RESIDENT

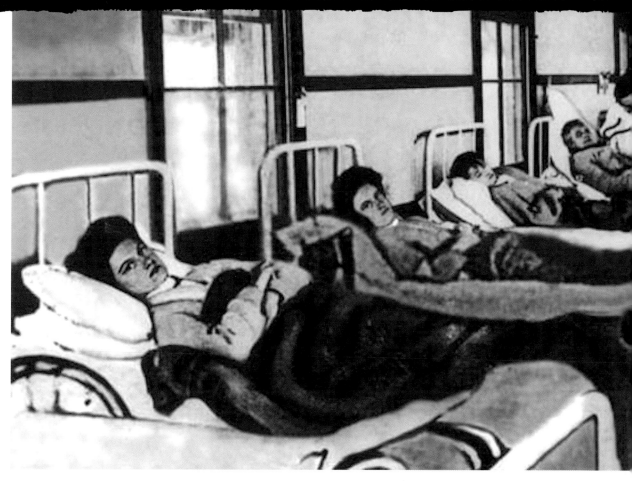

One of the island's most famous residents was Typhoid Mary. Mary Mallon emigrated from what is now Northern Ireland to New York City as a teenager. She found work as a private chef, cooking for elite families. Typhoid Mary became an example of why it's so critical that kitchen workers wash their hands well. The bacterium that causes typhoid fever is a form of salmonella (*Salmonella typhi*) that can live on in feces and urine, so a food preparer who did not wash his or her hands thoroughly could pass it onto food and then on to new victims.

As Mary went from one cooking job to the next, people began to fall ill with typhus. Oddly enough, Mary herself didn't have any effects from the disease, so at first it seemed unlikely that she was the carrier. After an intensive investigation, doctors discovered that she was an asymptomatic carrier: She harbored the deadly illness in her body but she never experienced any symptoms. The *San Jose Evening News* wrote in 1907: "Never has there been an

Mary Mallon, also known as Typhoid Mary, became famous for spreading deadly typhoid fever. Although she was the carrier, Mallon herself was immune to the disease. Public Domain

instance, as the present, where a woman who never had typhoid fever should prove a veritable germ factory."

In 1902, Mallon worked as a family chef in Maine at the summer home of a wealthy lawyer, Coleman Drayton. During her service there, seven out of nine people took to their beds with violent sickness. Because Mallon devoted much of her time to tending to the ill, Drayton never considered that she could be the source. In the summer of 1906, six out of eleven people staying at the summer home of New York banker Charles Henry Warren on Long Island were overcome with typhoid fever. After conducting a thorough investigation on his property and staff, he concluded that the disease could have come from only one source: Mary Mallon. Most household dishes were prepared hot enough to kill off bacteria, but Mary had served ice cream with fresh peaches cut up and frozen in it. Although it was a family favorite, the sweet treat was most likely the cause of their sickness. All in all, it's thought that Mallon infected about fifty-one individuals, and three of those died. As the public learned of her story, she gained the nickname Typhoid Mary.

Because Mallon resisted all calls for her to be voluntarily tested, New York City officials had her arrested in 1907 in the name of public health safety. They whisked her away to North Brother Island, where she lived in a small bungalow until 1910, all the while rejecting the notion that she could be a spring of contagion. Health officials released Mallon under the condition that she never work as a cook again. But strong-minded Mallon would hear none of that. Soon, she was back in the kitchen whipping up fresh cases of typhoid fever. On March 27, 1915, New York City Sanitary Police arrested her again. They dragged her back to North Brother Island where she was confined till her death in 1938, at age sixty-nine. In history, Mallon will always be remembered as Typhoid Mary, the personification of how a deadly epidemic can spread from just one individual.

A SHORT HISTORY OF QUARANTINING

The practice of isolating the sick so disease does not spread and make others ill is an ancient one. The Old Testament, for example, refers to the custom of isolating lepers. The word *quarantine* itself comes from fourteenth-century Italy—a time of the plague (aka, the Black Death). The plague (which was actually a germ called *Yersinia pestis*) claimed the lives of up to twenty-five million people across Europe, with some areas seeing a mortality rate of two-thirds of the population. Boats arriving in the busy port of Venice from regions known to be rampant with disease were required to anchor off the coast for forty days—or *quaranta giorni*—before docking. The phrase *quaranta giorni* evolved into "quarantine."

In 1348, Venice established the first quarantine to try and stop the spread of the Black Death. University of Vermont

England suffered a plague epidemic in 1665 that wiped out about 100,000 of London's 460,000 citizens. The city attempted to take quarantine measures to stop the spread of illness, but measures were haphazard. London was only saved from the plague by another deadly disaster—the Great Fire of 1666. The fire not only stopped the plague; it also destroyed almost the entire city as well. Cholera swept through New York City and London in the 1830s. As with Venice in the Middle Ages, ships in these cities were put under quarantine.

When the AIDS epidemic struck in America and Europe in the mid-1980s, many did not know how the disease spread. HIV, the human immunodeficiency virus that causes AIDS, does not survive well outside the body and is not spread by casual contact, but many thought otherwise. In 1985, the *Los Angeles Times* conducted a poll showing that a majority of Americans favored a quarantine of AIDS patients, and some embraced measures as drastic as using tattoos to mark those with the deadly disorder. As more was learned about the disease, these plans were never carried out, but even as recently as 2017, a Georgia state representative raised the idea of a quarantine as a way to stop the further spread of HIV.

In 2002, an outbreak of SARS (severe acute respiratory syndrome) began in Asia. A total of 8,098 people worldwide became sick, and 774 died. Beijing quarantined 30,000 residents. In Canada, 44 people died from SARS, and about 250 became ill. Toronto asked 25,000 residents to place themselves under self-quarantine in their homes. In 2005, China took quarantine measures to stop the spread of a deadly flu strain, H5N1, also known as bird flu. In 2015 when Kaci Hickox, a Doctors Without Borders nurse, returned to the United States after treating Ebola patients in Liberia, she was held in a quarantine tent in New

The *New-York Tribune* covered the devastating tragedy of the *General Slocum* disaster on June 15, 1904, when more than 1,021 died in the East River, many of them children. Library of Congress

DEVASTATING TRAGEDY
ON THE EAST RIVER

Three years before Mallon was imprisoned, North Brother Island was the location of one of the worst tragedies in New York City history. On the morning of June 15, 1904, the *General Slocum* riverboat departed from the Lower East Side filled with about 1,350 passengers and crew. The steamship was preparing to take a group of mostly women and children from Manhattan's Little Germany neighborhood on an end-of-school-year outing. At about 9:30 a.m., the ship began its journey up the East River. Its destination was Locust Grove, a picnic ground on Long Island's north shore. As the boat passed 97th Street, however, a fire broke out belowdecks. A panicked crew raced for the fire hoses on board, but found they had rotted and could not extinguish the blaze.

As the fire grew stronger, the captain directed the boat toward North Brother Island, where he hoped to beach the boat and evacuate all onboard. In the meantime, crew distributed life jackets, but most of these had rotted just like the fire hose. Although the boat came equipped with safety tools, none of them had been checked in ages. As flames swallowed more of the ship, passengers jumped into the water. Many of the women and children could not swim and drowned. Others burned to death on the ship. Still, Captain William Van Schaick somehow guided the *General Slocum* to within twenty-five feet of the island, where staff had raced to the shore with fire hoses. Although these rescuers were able to pull some survivors to safety, a total of 1,021 perished that day, making the burning of the *Slocum* New York City's worst disaster until the attack on the World Trade Center on September 11, 2001. The tragedy cast a pall on the German community in New York as well, and Little Germany—or *Kleindeutschland*, as it was called—swiftly declined and disappeared. A fountain memorializing those lost in the disaster stands today in Tompkins Square Park in the East Village in Manhattan.

Rescue workers carry the dead bodies from the *General Slocum* on the pier of North Brother Island. Public Domain

LASTING IMPACT:
FROM SANATORIUM TO BIRD SANCTUARY

The hospital on North Brother closed shortly after Mary Mallon's death. After World War II, the grounds changed into a home for war veterans. From the 1950s until 1963, the island served as a drug treatment facility for juvenile delinquents. After 1963, the island shut down. The power was cut off, and it has not seen any human inhabitants since. Under control of the New York City Parks Department since 2001, the island, with about twenty-six structures in various states of decay, now serves as a sanctuary for black-crowned night herons and other wading birds. In 2016, New York City Council members began discussing ways to reopen the island as a destination for the public, but as of now, North Brother remains deserted. ■

PART III

MOTHER NATURE STRIKES:
TOWNS LOST TO NATURAL DISASTER

Nothing much remains of the lumber town of Ruddock except a few signs on Interstate 55.
Photo by Scott M.X. Turner, used with permission

THE HEADLINE

RUDDOCK, NAPTON, AND FRENIER, LOUISIANA: WIPED OUT BY HURRICANE

Hurricane brings an end to a peaceful Southern lumber community

LOCATION:
Ruddock, Napton, and Frenier, Louisiana

DATELINE:
Early 1800s–September 29, 1915

On September 29, 1915, a Category 4 hurricane wiped out three small towns on the narrow Isthmus of Manchac in Louisiana. Today, Interstate 55 runs through the strip of land that separates Lake Pontchartrain and Lake Maurepas. Although a big green Exit 7 sign indicates that Ruddock lies ahead, nothing remains of the town but a boat launch. All that's left of the old Frenier community is a mass grave. Once-fertile farmlands remain submerged beneath the waters of Lake Pontchartrain. Environmentalists foresee that one day waters will rise to a point that the entire isthmus will be gone for good, and only one big lake will remain. Many travelers just pass through the area now as they head to the next nearest towns of either Ponchatoula to the north, or Laplace to the south.

ISTHMUS: A narrow strip of land, bordered on both sides by water, connecting two larger bodies of land.

WHAT DO WE KNOW?

At the turn of the century, the economies of Ruddock, Napton, and Frenier thrived along the Manchac swamp. Photo by Scott M.X. Turner, used with permission

The three towns, all part of St. John the Baptist's Parish, were originally settled by German immigrants in the early 1800s. When the railroad came through in 1854, it opened the door to more commerce. Before the train tracks were built, townspeople shipped goods in and out on boats that traveled across the lake.

In 1892, William Burton and C. H. Ruddock established the Ruddock Cypress Company, taking advantage of the railroad for lumber transportation. The lumber company erected its buildings on wooden stilts rising out of the swamp waters. Wooden sidewalks were also constructed to keep locals from trudging through the muck.

"The houses at Ruddock were on the east side of the tracks and were all two-story homes," wrote Wayne Norwood in his book, *The Day Time Stood Still: The Hurricane of 1915.* "There was a main walkway down the middle of the street, which was a slab of cypress cut and laid in the swampy area. Houses were on each side of the walkway, and a community center was located at the end."

The entire town's economy centered around the sawmill, with workers harvesting large cypress trees from the Louisiana swamp using crosscut saws and axes, and floating or dragging the timber through the swamp to the Ruddock mill. Although it was backbreaking work in conditions that were often hot, humid, and filled with mosquitoes, the community thrived and quickly swelled to a population of about one thousand.

"Life may have been hard back then, but we loved where we lived in the swamp," said Mrs. Helen Schlosser Burg, a longtime resident who was interviewed by Norwood.

Born in 1901, Burg grew up in the even smaller town of Wagram, just about five miles south of Ruddock along the train lines. The town was later renamed Napton, supposedly because residents in the area enjoyed a good afternoon snooze from about one to three p.m. A bit further south from there along the track was the town of Frenier.

Some Frenier and Napton residents worked at the sawmill, but they also farmed cabbage and manufactured barrel staves. They would pack the cabbages in locally made barrels and ship them out along the rail lines. Along with a diet heavy on cabbage, residents hunted game like rabbit and deer, and ate seafood from the lakes. When folks needed to supplement their local supplies, they would hand a grocery list to the train engineer as he passed through. He would purchase goods in New Orleans and have them ready for pickup on his way back. Folks here lived cheaply; Mrs. Burg said that a typical grocery bill for the year would be about $100.

After the turn of the century, the towns slowly crept into the modern world, establishing a post office, telegraph office, and boardinghouse. Still, many homes had no electricity, and water came from cisterns built to catch rainwater. Land sold for just twenty-five cents an acre. There were no cars or police in this peaceful, sleepy region. People left their doors unlocked. Families relied on home remedies to treat sickness, and babies were delivered by midwives. For a real doctor, people had to travel to a bigger city. Residents went every Sunday to the Holy Cross Catholic Church, but the priest came through only once a month. The Catholic school held classes sporadically when a nun could make the trip.

For entertainment, male laborers blew off steam in the male-only Owl Saloon; other townspeople would attend a once-a-month house party, which neighbors would take turns hosting. They'd cut loose from late afternoon Saturday until Sunday morning, performing music, singing, dancing, and playing games. Some of the boys would get dressed up, slicking back their hair with alligator oil.

TIMELINE:
RUDDOCK, NAPTON, AND FRENIER, LOUISIANA

Early 1800s	German immigrants settle in the region
1854	Railroad opens on the Isthmus of Manchac
1892	Ruddock Cypress Company opens
Early 1900s	Area gets a telegraph office and post office
1915, September 29	The Great New Orleans Hurricane hits

TROPICAL HURRICANE DEVASTATES THE SOUTH

Five hundred persons are believed to have lost their lives in the tropical hurricane that raged over the lower Mississippi valley. The extent of the tremendous damage wrought has not yet been fully calculated, but probably the figure will be well over $12,000,000. The upper photograph shows the wreckage of the railway depot and St. John's Rowing club buildings at West End, on Lake Pontchartrain near New Orleans. The lower photograph shows the wreck of a ferry boat, barge and gasoline boat at the Tugger landing on the Mississippi river.

The hurricane obliterated hundreds of buildings in its path, including the cupola of the First Presbyterian Church in in New Orleans' Jackson Square, which smashed to the ground. Courtesy of *Madison Journal*, Public Domain

THE EVIDENCE

The hurricane began forming in the Caribbean on September 25, 1915. In her interview, Mrs. Burg, who was fourteen when the storm swept in, described how her father read news of the approaching bad weather: "One day, Daddy got the newspaper from the train and it told of a hurricane. Daddy said, 'You know, one day we are going to get a bad storm here.' We went to bed that night; we had no way of knowing that our lives were about to change forever, and for some, this would be the end."

This map shows the path of the devastating hurricane of 1915. Image created using WikiProject Tropical cyclones/Tracks; background image from NASA; tracking data from National Hurricane Center

As the New Orleans Hurricane of 1915 (also called the Great West Indian Hurricane) headed toward land, its winds reached as high as 145 miles per hour. They weakened to about 120 to 129 mph when they hit land on September 29, but the winds were still a deadly force. All in all, the hurricane killed at least 275 people and caused $13 million in damage (about $315 million in today's dollars).

In New Orleans, the hurricane ripped the roofs off buildings, including the cupola of First Presbyterian Church in Jackson Square, which was sent flying to the ground. Eleven steeples in all were demolished. Many structures were flattened.

In Ruddock and the other small towns on the east side of the railroad tracks in St. John's Parish, the heavy winds and rain swept in at night and kept going throughout the morning. "The rain was so strong, it was just like buckshot hitting you in the face," Burg said. "Them waves was almost hitting under the house when Papa took us out."

The Waters Kept Rising

Waters rose as high as thirteen feet. Fearing for their lives, Burg's father put two of the children in his small pirogue (a long narrow canoe made from a single tree trunk). He rowed it to the railroad tracks, which were still higher than the encroaching waters. After getting the kids onto the tracks and giving them a tablecloth to shield them, her father tried to row against the fierce winds to retrieve the rest of the family. The journey was a struggle, as the wind kept blowing him back. With superhuman effort, he finally got the whole family together and they slowly made their way to the schoolhouse on the other side of the tracks.

Soon, floodwaters found their way to the other side of the tracks as well, and waves began crashing up against the side of the schoolhouse. The powerful winds made it feel as if the whole structure might blow away, so Burg's father again retrieved his small boat and squeezed the entire family into it. Swimming beside it, he dragged his family farther west.

Just as the family's situation was looking most dire, they heard a sound of rescue—the whistle from Train No. 99. Somehow the tracks weren't completely flooded over, and the train made it through to take residents south to New Orleans. It was already full of evacuees from the isthmus. Shouting to the train, the Burg family frantically paddled with their hands. As they passed back by the school, they saw that it had blown over. The Burgs scrambled aboard the train, happy to get some shelter from the storm. As they chugged off, they breathed a sigh a relief. But after just a few miles, the train stopped. A patch of track ahead was washed away and covered with debris. The conductor decided to put the train in reverse and make their way north toward Manchac. The train rolled on a short distance and lurched to a stop. The track had been washed out in that direction as well. Everyone onboard gathered together in the caboose. Soon water was rising within the train car as waves slapped against the outside. Passengers dropped to their knees in the water and started to pray. Outside, survivors clung to cypress trees, screaming for help, as others drowned in the flood.

Wiped out in 1915, the town of Frenier has bounced back—somewhat. The welcome sign now boasts 38 residents. Photo by Scott M.X. Turner

DEVELOPING STORY:
AFTER THE STORM

Although New Orleans suffered millions of dollars in damage in the hurricane that struck on September 29, 1915, the city survived. Ruddock, Napton, and Frenier were not so lucky.
Public Domain

All in all, about fifty-eight residents were killed in Ruddock, Napton, and Frenier. By the morning of October 1, the winds had died down and the waters started to recede. The storm had been so powerful that only one home was left standing. All other buildings had been blown over and washed away. Ruddock, Napton, and Frenier were completely gone. Survivors discovered bodies among the debris, in the woods, and in the swamp. Two residents who had been out of town when the storm hit returned to erect a mass graveyard where the victims were all buried, and shortly afterwards these small towns gave up. The devastation was complete, and no one wanted to live through such a disaster again.

Residents of New Orleans rode through the flooded streets in a high-wheeled wagon to "see the sights." Public Domain

FRINGE THEORY:
DID JULIA BROWN CURSE THE TOWN?

Legend says that Julia Brown, resident of the nearby Manchac swamp, foretold the disaster. The Manchac swamp lies to the north of Ruddock, filled with insects and alligators. Here, Julia Brown would sit on her porch and give all who passed the evil eye. She was known as a voodoo priestess who doled out charms and curses. She would often sing a haunting tune: One day I'm gonna die, and I'm gonna take all of you with me. Some say that because the town took her for granted and thought she was a kook, she was seeking her revenge with this curse.

*One day I'm gonna die,
and I'm gonna take all of
you with me.*

A sign on the way into Frenier remembers Julia Brown, who lived on the edge of the swamps along Lake Pontchartrain. She sat on her porch singing songs, including one where she warned that when she died, she would take everyone with her. Photo by Scott M.X. Turner.

Julia Brown died just before the hurricane. As the townspeople were holding her funeral, the hurricane swept in. Supposedly, as Brown's coffin floated out into the swamp, the three towns were obliterated. When the waters receded, they found Julia's body deep in the swamp, but her casket had disappeared. Several visitors to the Manchac swamp claim they have experienced her ghost, hearing the screams of a woman or a disembodied voice singing the song of Julia Brown. Census records show that there actually was a Julia Brown who lived in the area at the time. A modern voodoo priestess believes Brown was most likely a local faith healer (called a *traiteur*) who would help the sick and perform midwifery duties. She believes that Julia Brown's song was more of a warning than a curse. But to this day, many Louisiana locals believe Julia's black juju brought an end to Ruddock, Frenier, and Napton. ∎

GONE FOR GOOD: OTHER US TOWNS DESTROYED BY HURRICANES

TOP: What remains of Dernière Island (or "Last Island") helps to provide Louisiana coastal towns with a storm barrier. Newspapers.com

The newspaper story at bottom recounts one survivor's tale of the Isle Dernière Hurricane. Newspapers.com

Severe weather has caused the demise of several towns in the United States. Here are a few that were wiped out by wild weather:

Isle Dernière, Louisiana

Located just five miles south of the mainland in the Gulf of Mexico, Isle Dernière (meaning "Last Island"), with its beaches and cool breezes, beckoned the wealthy of Louisiana in the first half of the nineteenth century. During the 1840s, the well-to-do built summer homes there, and visitors arrived by steamboat to enjoy ocean swimming and horse and carriage rides. Vacationers could take a spin on the carousel, or bowl or play billiards at the sprawling Muggah's Hotel. For years, the resort town remained a place to forget your worries and have a carefree trip.

But that all changed on August 10, 1856. Although residents had experienced rainstorms before, they'd never seen anything as devastating as the hurricane that would sweep through at the end of that weekend. On the Friday before the storm arrived, inhabitants noted that the ocean waters looked angry and waves were growing higher. By the end of the weekend, torrential rains were pummeling the island, with winds that revved up as high as 150 miles per hour. When a tidal wave came crashing in, as many as 150 people were swept out to sea. The *Star* steamboat ferry provided shelter for about 160 who were able to weather the battering. All homes on the island were obliterated. A total of 200 people died that day, and after this storm, no one would ever return to inhabit the island again. Because of rising waters, Isle Dernière is now a string of five smaller islands, its main residents now pelicans and other seabirds.

Indianola, Texas

Located on Matagorda Bay in the Gulf of Mexico, Indianola was a thriving port second only to Galveston when it came to Texas shipping at the end of the nineteenth century. German immigrants built the city in the 1840s. In September 1875, 5,000 residents were living there when a major hurricane ripped through. Although 176 died and about three-fourths of the town was washed away by floodwaters and fierce winds, the remaining residents were determined to get back on their feet. They rebuilt the town largely as a resort destination for fishing, swimming, and dining. But then during the summer of 1886, a Category 4 hurricane blasted Indianola again. On August 20, powerful winds razed the rebuilt structures. A lamp blown off in the winds ignited a fire, which spread through many of the buildings. Overnight, the seaport town was gone, never to return.

In 1886, newspapers reported on the violent storm that destroyed Indianola. Newspapers.com

Located on Matagorda Bay in the Gulf of Mexico, Indianola was a thriving port city. Built by German immigrants in the 1840s, it was soon Texas' second-largest shipping center, next to Galveston. Courtesy of the Calhoun County Museum, Port Lavaca, Texas

Hog Island, New York

Near Rockaway Beach in Queens, New York, Hog Island supposedly got its name because its shape resembled the back of a pig. Others say that Native Americans raised pigs there that were brought by Europeans. Just a mile long and a couple hundred feet wide, the island was actually a big sandbar.

After the Civil War, the island became a popular destination, attracting swarms of city dwellers for its beaches, bathhouses, saloons, and two or three restaurants that provided food and entertainment. According to an article in the *New York Times* in 1874, "The majority of the visitors [to Far Rockaway] spend almost all their time on the island, enjoying the cool ocean breezes to be found there, even in the most extreme heat." Bellot's *History of the Rockaways*, published in 1917, recounts how New York City politicians from Tammany Hall (the Democratic political organization) met frequently at a Hog Island restaurant owned by Patrick Craig to discuss strategies, campaigns, and various deals. In the 1870s, Irish officials controlled Tammany Hall, and Hog's Island gained the nickname of the Irish Saratoga. Nights were filled with oysters, steaks, liquor, and cigars.

On the night of August 22, 1893, as strong waves beat against the shore, news started coming in that a Category 2 hurricane was heading up the coast toward New York City. Fortunately, August 23 would be a Wednesday and not many would be visiting the island. The skies began to darken by late afternoon, and at eight p.m., the storm hit. In Central Park, trees were uprooted. The East River crested the seawall, flooding the streets of Brooklyn and Queens. A few unfortunate police horses got caught in torrents of seawater and drowned. As the ferocious hurricane lashed Hog Island, restaurants and casinos crumbled, and those on the island fled via rowboats to Rockaway.

The Friday edition of the *New York Times* reported that "hundreds of chimneys were tossed down like playthings, and roofs were ripped off as if with a knife." Telegraph wires fell "like cotton strings." The next day, only the "unbroken surface of water" could be seen where thousands of pleasure seekers had visited and thousands of dollars had been invested. Luckily, no one died. The sandbar slowly returned and saw life again as a swimming destination. But after another storm in the spring of 1902, Hog Island was gone for good. Today, however, beachcombers still find remnants of the former resort destination as pieces of plates and beer mugs wash up on nearby Rockaway Beach. ∎

Hog Island was similar to nearby Rockaway Beach, offering city dwellers a seaside escape for ocean swimming during the hot summer months. Public Domain

THE HEADLINE

VANPORT: FLOOD WASHES AWAY OREGON'S SECOND-LARGEST CITY

The country's biggest wartime housing development vanishes in a day

LOCATION:
Vanport, Oregon

DATELINE:
1942–48

In the late spring of 1948, the Columbia River overflowed its banks and broke through the dikes protecting Vanport, Oregon, a city located on the northern side of Portland. The community had been built by Portland officials to accommodate the large number of workers needed to complete US Navy ships for World War II. At its peak, at least thirty-five thousand lived there. In a region that had a predominantly white population, both black and white residents lived in Vanport. Although segregation was common, there were integrated schools and community centers. The first black teachers and policemen in Oregon were hired in Vanport. By 1948, with the shipyards closed, the number of residents in Vanport dropped to eighteen thousand, as many moved out to seek better fortunes. A large percentage who stayed on in the city were African Americans. The flood that year marked the end of Vanport, but its history played a crucial role in race relations and the development of modern-day Portland.

TIMELINE

c. 1910   Albina district in Portland becomes an African-American neighborhood

1938   Bonneville Dam completed

1940   Henry Kaiser builds three naval shipyards

1941   The United States enters World War II

1942   Vanport city is erected to house influx of workers. Population of city reaches thirty-five thousand

1945   World War II ends

1947   Vanport population declines to eighteen thousand

1948   Flood wipes out all traces of Vanport

A boy clings to safety at Entrance Circle during the Vanport flood on Memorial Day, **1948.** Barcroft, photo by John Killen

THE EVIDENCE:
DEALING WITH A HISTORY OF SEGREGATION

P ortland is considered by many to be a progressive and liberal city. The television show *Portlandia* satirized how very "politically correct" the residents can be. The program also highlighted the city's lack of diversity, as only 6 percent of its population of six hundred thousand are African-American. When Oregon became a state in 1859, its constitution prohibited African Americans from living, working, or owning property there. After the Civil War, many whites from the South moved to the region, and the Ku Klux Klan thrived there in the early 1900s. During the early 1920s, the "Invisible Empire" (or KKK) influenced virtually every part of the state. In 1922, the KKK had the political power to get their pick for governor—Walter Pierce—elected.

As dominating as the KKK was, people were also challenging the organization, accusing the group of social and political divisiveness. In the mid-1920s, a new state law allowed African Americans to move into the state, but districting codes established

Denver Avenue is swallowed by floodwaters from the Columbia River after a major dike collapsed. Barcroft, photo by John Killen

by Portland bankers and realtors blocked blacks from living in certain neighborhoods. The city became known as one of the most segregated in the Northwest. Before World War II, most African Americans lived in a two-mile-long, one-mile-wide Portland neighborhood called Albina. After the Civil War, many whites from the South had moved to the region, and the Ku Klux Klan thrived there in the early 1900s. During the early 1920s, the "Invisible Empire" (or KKK) influenced virtually every part of the state. In 1922, the KKK had the political power to get their pick for governor—Walter Pierce—elected.

DEVELOPING STORY:
AN INFLUX OF DIVERSITY

In 1940, as the United States was preparing to possibly join the Allies in World War II, industrialist Henry Kaiser agreed to set up shipyards in Portland to supply the British Navy with vessels. When the Bonneville Dam opened forty miles east of Portland, it gave the region a plentiful source of cheap electricity. With two major shipyards in Portland and one in nearby Vancouver, Kaiser constructed 752 ships during the war. The booming industry fueled population growth as workers migrated from around the country, including thousands of African Americans. During the 1940s, an estimated 1.5 million African Americans left the South for better opportunities in the North. Before the war, Portland was home to about 1,800 African Americans, but that num-

ROSIE THE RIVETER

"*Rosie the Riveter*" was a cultural icon of World War II, a character representing the women who worked in factories and shipyards, like the ones in Portland. Her image was used to help recruit female workers for the war effort as the labor force dwindled because men were joining the military. J. Howard Miller, who created the character, likely based her on Naomi Parker Fraley, who worked in a navy machine shop in California. Fraley died on January 20, 2018, at the age of ninety-six.

We Can Do It!

ber steadily climbed up to 15,000 by 1946. Shipyard work also attracted nearly 40,000 Native Americans, and offered opportunities for thousands of women.

In 1942, to house the enormous influx of workers, Kaiser erected a new city on the northern side of Portland, in marshland that was kept dry by dikes along the Columbia River. First nicknamed "Kaiserville," the city officially became Vanport, a name taken from the cities on either side of the river—Vancouver and Portland. More than ten thousand homes and apartments were constructed here in just 110 days. Most of the homes on these 648 acres of swampland were hurriedly constructed of wood, on wooden foundations. Residents often felt penned in—surrounded by fifteen- to twenty-five-foot-high dikes on all sides, cutting off any view of a horizon.

Almost overnight, Vanport (or "The Miracle City") became the second-largest city in Oregon, with thirty-five thousand residents, many of them African-American. At the time, it was the largest housing project in America. Kaiser established social services and a health-care system for workers, which became the foundation of the Kaiser Permanente health-care system. The metropolis boasted its own movie theater, library, police station, hospital, recreation centers, playgrounds, and schools, all of which were integrated. Vanport schools became the first in Oregon to hire African-American teachers.

During World War II, shipyards like those in Portland depended on female workers to help build the vessels America needed to fight Nazi Germany. The character Rosie the Riveter was used to help recruit women to work in factories. By J. Howard Miller (1918–2004), artist employed by Westinghouse, poster used by the War Production Co-ordinating Committee

THE TURNING POINT:
THE DAY THE DIKE BROKE

When the war ended, shipbuilding died out as well. Subsequently, the population swiftly dropped to eighteen thousand, or just about half of what it had been at its peak. Many former workers suffered from severe health problems because of exposure to asbestos. Along with unemployed shipbuilders, veterans and others with low incomes, who could not afford better housing, remained. About one-third of Vanport's postwar population was African-American. City historians have said that most had few other options because of Portland's discriminatory housing policies. The city began eyeing Vanport as a location for industrial redevelopment, and talk began on how Vanport might one day be dismantled.

ABOVE: **After the flood, Vanport homes were uprooted, bobbing in the water like toys in a large bathtub.** Barcroft, photo by John Killen

LEFT: **Residents form a human chain to rescue those trapped in the floodwaters.** Courtesy of Oregon Historical Society Library

FRINGE THEORY

Some Vanport residents believed that city officials did not warn people as soon as they could have because Vanport was held in low regard. Also, some suspected that the dikes were not fortified as well as they could have been because the city wanted the town to be destroyed to make way for new development.

Local citizens like Carl Downey (pictured, center), rescued Vanport residents. Barcroft, photo by John Killen

Vanport gained a reputation for being a seedy neighborhood, although the crime rate wasn't any higher than that of the rest of Portland. According to *The Oregon Encyclopedia*, Mayor Earl Riley called it a "great headache" and a "municipal monstrosity." A 1947 report in the *Oregon Journal* called Vanport an eyesore. As city officials and developers debated the future of Vanport, Mother Nature made the decision for them.

The winter of 1948 brought particularly heavy snows to the mountains around Oregon. Warm spring weather combined with steady May rains filled the Columbia River to dangerous heights. As the river surged, Portland authorities saw potential trouble coming. On May 25, officials were sent to patrol the dikes.

By May 28, the Friday of Memorial Day weekend, the waters had risen to about twenty-five and a half feet, and weather forecasters were calling for levels to rise to thirty feet, or higher.

The Portland Housing Authority had been keeping a watchful eye on Vanport, and had determined that the dikes would hold. On Sunday, May 30, Housing Authority employees slipped notices under Vanport residents' doors, telling them that the dikes were secure. If the situation changed, sirens and air horns would blare, but the message read: "Barring unforeseen developments, Vanport is safe. . . . You will be warned if necessary. You will have time to leave. Don't get excited."

As the day progressed, temperatures grew warmer, hitting a high of about 76 degrees. Many people spent the afternoon outside, enjoying the warm weather. On the Westside, a dike built from an old railroad trestle was thought to be the strongest, but it actually was constructed of just fill dirt and wood. At 4:17 p.m., the dike collapsed. Within minutes, sirens were blaring. All eighteen thousand residents had only about thirty-five minutes to evacuate.

Cars are overturned in the floodwaters at Vanport, Oregon, in 1948. Portland Housing Authority, Public Domain

Buildings Were Swept Away in a Flash

The break in the big dike quickly stretched to sixty feet, and then to five hundred feet. Ten-foot-high waves rushed into the city, washing most of Vanport's apartment buildings from their wooden foundations. Cars tumbled and bobbed in the rising flood.

In an interview on Oregonlive.com, Ed Washington, who was a youngster at the time, recalled how his mother and five siblings rushed to gather a few things in their home. When they left, there was some water seepage; Washington thought they'd be able to return home later and just sweep the water out. But when they got to Denver Avenue, he looked back to see a wall of water lift his home and others off their foundations, crashing them into each other and leaving them in utter ruins. Entire structures, including Vanport's firehouse, were swept up in the fierce currents. Electrical lines snapped and sizzled in the water. As the waters reached shoulder-high, people scurried to the tops of embankments and created human chains to pull each other to safety. Within an hour, Vanport had turned into a large lake, and by the end of the day the entire city was gone. A total of fifteen residents were killed.

LASTING IMPACT:
A MOVEMENT TO IMPROVE CONDITIONS

The displaced population heightened racial tensions. At first, citizens put aside prejudices as whites took blacks into their homes, and vice versa. But in time, black families had to find permanent housing, and many were forced to make homes in the already overcrowded Albina district. Today, Portland remains largely white, but has moved over the years to dismantle racist policies and become less discriminatory. The historically black neighborhood of Albina, however, has become increasingly gentrified, and the quest to find fair and decent housing for those of all incomes and backgrounds remains a struggle.

To minimize future flooding problems, the US Army Corps of Engineers developed a multiuse reservoir storage plan for the Colum-

bia River Basin. A 1964 treaty with Canada led to the development of millions of acreage of water storage for flood control. In 1996, Portland again faced severe flooding, although dam operations helped to reduce the damage they caused. North Portland also minimizes flooding with a system that includes a combination of levees (dikes) and sloughs (drainage channels) near the Columbia River. Some forty-five miles of levees reduce the risk of flooding from the Columbia River, while the sloughs help to drain water from the land.

Today, the land where Vanport once stood is home to Portland International Raceway, Heron Lakes Golf Course, and West Delta Park. ■

Mr. and Mrs. Leonard C. Davis didn't let racial discrimination affect their decision to open their home to victims of the Vanport flood. Here they chat over coffee with Ms. Bertha Freeman, who was displaced by the disaster.
Barcroft, photo by John Killen

OTHER DROWNED AMERICAN TOWNS

Old Cahawba, Alabama

labama's first state capital, Old Cahawba, became a ghost town by the early 1900s after a series of floods. After Alabama became a state in December of 1819, the capitol was built at the spot where the Alabama and Cahaba rivers meet. Using the city of Philadelphia as a model, Cahawba became a bustling center of political and social life for the state. The *Encyclopedia of Alabama* writes that the metropolis soon boasted numerous stores, a state bank, several hotels, two ferries, several physicians, eight lawyers, and two newspapers. The town was not without its hardships; the river often swelled beyond its banks, and a hardy mosquito population spread disease. In 1826, the Alabama State Assembly voted to move the capital to Tuscaloosa.

Cahawba, the original capital of Alabama, was abandoned because of constant flooding. Newspapers.com

Still, Cahawba thrived. As cotton plantations in the region grew, the city served as an important commercial hub, with paddlewheel steamships busily working its port. Completion of the railroad in 1859 brought thousands more to the region. During the Civil War, the Confederacy established a huge prison there that housed more than three thousand Union soldiers.

After a big flood destroyed much of the city in 1865, and the town lost it railroad terminal, Cahawba began to decline. By 1870, only three hundred residents remained. By 1900, it was abandoned. All that remained were the cemeteries, ruins, and a few old buildings.

These brick columns are all that remain of the Crocheron Mansion, built in 1843 in the town of Cahawba, Alabama. Old Cahawba, Alabama Historical Commission

As with many ghost towns, Cahawba has a resident ghost: An apparition described as a large, white, luminous floating ball has appeared before several visitors. It became known as Pegues's Ghost, because it showed itself near the home of Colonel C. C. Pegues, who had been mortally wounded in battle. More scientifically minded researchers think the ghost may be a phosphorescent phenomenon known as Will-o'-the-Wisp, an atmospheric occurrence that appears over swamps and bogs, thought to be caused by methane bubbles that rise to the surface and burn spontaneously in the air.

The Catskills Underwater

To quench the thirst of its millions of residents, New York City built a series of aqueducts that connect to reservoirs in upstate and western New York, and deliver billions of gallons of water to the Big Apple each day. For the most part, the water travels via gravity, as is the case with the Catskill Aqueduct, which stretches about ninety-two miles from the Ashokan Reservoir in the Catskill Mountains to the northern boundary of the city. Completed in the early 1900s, the Catskill Aqueduct is an astounding feat of engineering. To form the Ashokan Reservoir, which supplies about 40 percent of New York City's water, dozens of towns had to be flooded.

Catskill Aqueduct under construction, 1911.
Wikimediacommons.org.

To build a reservoir big enough to supply the city, the New York City Board of Water Supply needed to acquire a dozen nearby towns, forcing the two thousand people living in these hamlets to move. Although some residents wanted to stay, the city and state forced residents to move, exercising the right of eminent domain. The city bought thousands of acres, much of it farmland, although many residents of the Ashokan region said that the prices were not fair.

Eight towns were moved, while four others were burned to the ground. The destruction went beyond homes; churches, schools, businesses, doctors' offices, blacksmith shops, sawmills, and railroad stations were all removed. The city even offered residents $15 each to dig up dead relatives so that cemeteries could be relocated. Many families had lived in these towns

WATERSHED: A watershed is an area of land that drains all the streams and rainfall to a common outlet, such as the outflow of a reservoir, according to the US Geological Survey.

EMINENT DOMAIN: The right of a government to take private property for the public good, and in return provide payment to the owners.

for generations, so it was incredibly difficult to think of them being erased from the map. In the 2002 documentary Deep Water, a woman whose family lived in the area said, "I'm sure they [my grandparents] were heartbroken, losing a twelve-room home and their store and all. My
mother had fond memories of that area as a child . . . and never really got over . . . the hurt of losing the land, home, friends."

Construction of the reservoir began on September 5, 1907, and the dam and dikes system was completed in 1914. Over the next year, the two Ashokan basins were filled with nearly 123 billion gallons of water.

In the 1950s, the state reservoir system grew with the completion of the Rondout, Pepacton, and Allegheny reservoirs, leading to the destruction of nearly ten more towns. Ironically, the town of Neversink was sunk below the water of the Neversink Reservoir in 1955. Workers completed the final reservoir in 1964, eliminating the town of Cannonsville, which was founded in the late 1700s.

The village of Shokan in 1906, one of the eight towns destroyed to create Ashokan Reservoir. New York Public Library Digital Collections Photo by Paul Van der Werf

PART IV

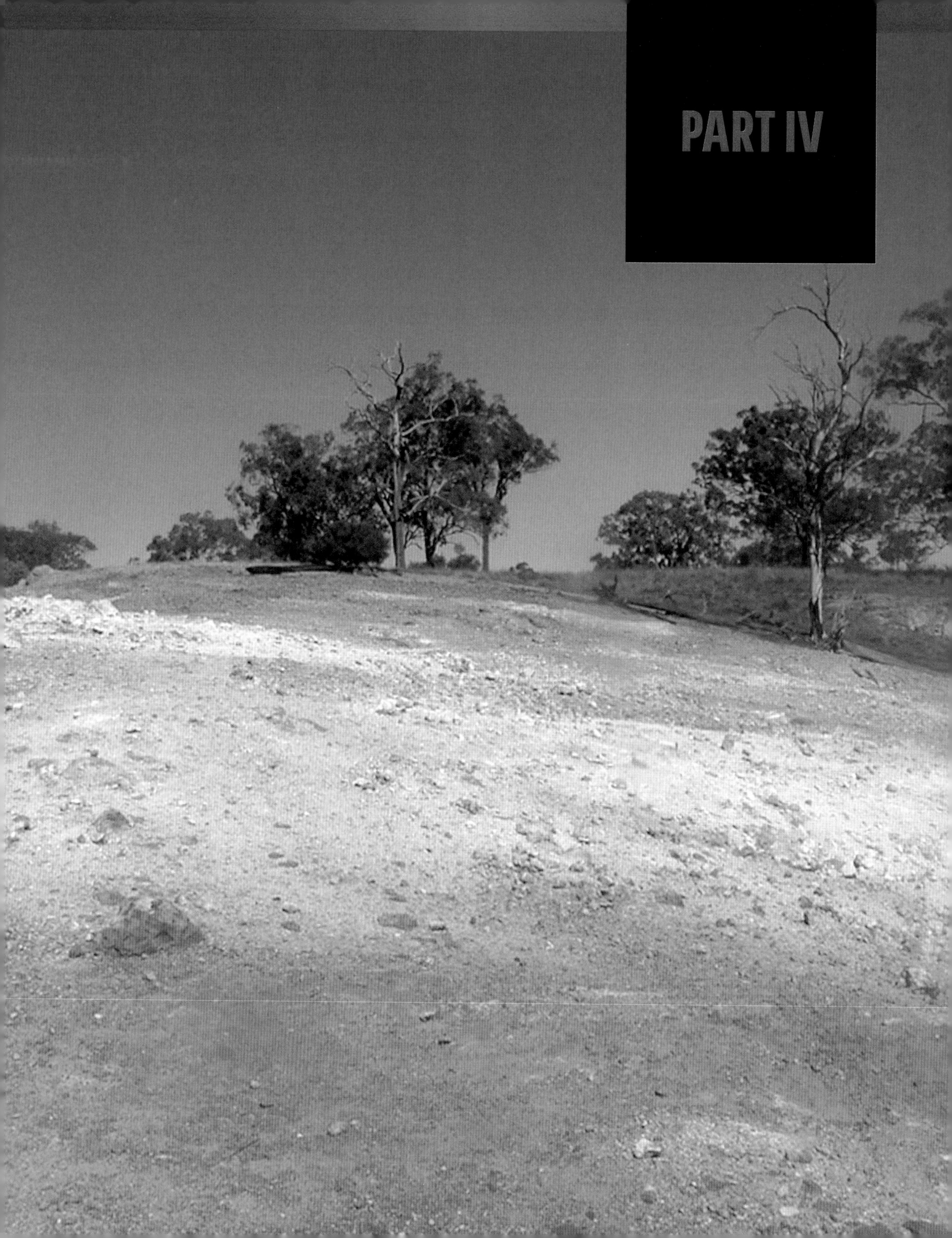

MAN-MADE DISASTERS: TOWNS TOO POLLUTED TO SURVIVE

THE HEADLINE

LOVE CANAL: AMERICA'S FIRST TOXIC GHOST TOWN

The poisoning of Love Canal led to a massive environmental cleanup

LOCATION:
Love Canal, New York

DATELINE:
1955 to Present

> *Give Me Liberty.*
> *I've Already Got Death.*

—From a sign posted by a
Love Canal resident in 1978

From the 1950s through the 1970s, Love Canal was a pleasant, working-class town situated about four miles from Niagara Falls. Over the years, an increasing number of residents suffered health problems, from dizziness and rashes to cancer, miscarriages, and birth defects. Most who lived in this peaceful enclave had no idea that their neighborhood sat atop more than twenty thousand tons of highly toxic chemicals, and that these poisons were seeping into the soil, water, and atmosphere. Although Love Canal became one of the biggest environmental tragedies in US history, it led to a movement that helped clean up hundreds of hazardous waste sites across the country.

In 1978, adults and children marched the streets in protest of the poisons contaminating their community in Love Canal. Newspapers.com

In June 1980, President Jimmy Carter and New York officials were burned in effigy because of the federal government's response to the toxic chemical pollution at Love Canal, where many residents suffered from a range of troubling symptoms. Newspapers.com

WHAT DO WE KNOW?

O riginally intended to be a dream community, Love Canal got its name from entrepreneur William T. Love, who arrived in Niagara Falls, New York, in the 1890s with a vision to create a utopian metropolis. Seeing that Niagara Falls and surrounding waterways could generate vast amounts of hydroelectricity, Love predicted that businesses would flock to the region to take advantage of this powerful source of energy. He developed a plan to build a model industrial city along a canal linking the Niagara River with Lake Ontario.

Love began construction, but the work of inventor Nikola Tesla would have a disastrous impact on Love's plans. Before Tesla, Thomas Edison had been constructing generators that produced direct current electricity that needed to be close to the power source, as the DC voltage could not travel very far without losing energy. Tesla's alternate current system, however, was excellent for long-distance transmission. With Tesla's AC generators gaining in popularity, the incentive for businesses to move near Niagara Falls evaporated. This factor, combined with a serious economic depression from 1893 to 1897, led Love to abandon his utopian dream, leaving behind a partially dug canal of about a mile.

Love sold the pit to the city of Niagara Falls, which began using it as a municipal chemical dumpsite. Between 1942 and 1953, Hooker Chemical Company took over the land and used it to dispose of about twenty-one thousand tons of toxic chemicals. These included at least twelve known carcinogens, including halogenated organics, chlorobenzenes, and dioxin. After filling the canal with immense amounts of poisonous material, Hooker Chemical covered the pit with dirt and clay, then sold the property to the Niagara Falls School Board for $1. Although the deal included a warning about the chemical wastes buried on the land, and a clause absolving Hooker Chemical of any future liability, the new community of Love Canal forged ahead with little concern over the fact they were sitting on a poisonous time bomb.

In 1955, the Board of Education constructed the 99th Street Elementary School near the old canal, opening its doors to four hundred students. Around the same time, hundreds of homes sprang up on the property. Love Canal quickly grew to become a solid working-class neighborhood.

TIMELINE

1890s William Love starts work to develop a utopian community near Niagara Falls

1942 Hooker Chemical Company begins dumping chemical waste into Love Canal

1953 Hooker Chemical fills in the canal and sells the property to Niagara Falls Board of Education

1955 The 99th Street School is finished, and hundreds of homes are built in Love Canal

1976 Heavy rains and snowfall lead to exposure of chemical barrels and widespread contamination

1978 President Jimmy Carter declares a state of emergency; residents are evacuated from seventy acres designated as the contamination zone, and 239 homes are razed

1980 President Carter declares a second federal emergency; about seven hundred more families move out of Love Canal. The EPA creates the Superfund program to clean up the nation's worst hazardous waste sites

1990 The head of the Environmental Protection Agency announces that much of the Love Canal neighborhood is safe enough for people to move back in

THE EVIDENCE:
SERIOUS ILLNESS SPREADS

In 1976, heavy rains and oversaturation from a record-breaking blizzard caused rusted barrels of waste to work their way to the ground's surface. Chemical-laden moisture seeped into basements, and puddles of oozing slime dotted backyards and playgrounds. In some cellars, black and red liquids seeped through the walls. Some of the children came home with mysterious burns on their hands and faces.

That year, as more and more residents complained about sickening odors in their yards, the city and county hired the Calspan Corporation to investigate. The firm detected poisonous chemical residues in the air and in the sump pumps of many homes, as well as high levels of polychlorinated biphenyls in the storm sewer system. Although Calspan advised that intensive measures be taken to seal off home sump pumps and contain the migration of wastes, the city did nothing more than provide a few window fans to ventilate those homes with high contaminant readings.

Increasingly, the residents complained of health issues—fatigue, irritable moods, red eyes, headaches, breathing difficulties, weight loss, dizziness, and rashes. People were missing work. One woman had epileptic-like seizures. Cancer cases inched up. Pets also suffered, losing fur and developing sores or tumors. One family described how fresh paint on their house was peeling off. When the New York State Department of Health (NYSDOH) conducted a health and environmental study in 1978, it found that a high percentage of women were reporting reproductive problems; the number of miscarriages and children born with birth defects was rising. Four babies were born with clubfeet, deafness, retardation, or other health problems.

On August 7, 1978, US president Jimmy Carter announced a federal health emergency at Love Canal. It marked the first time in US history that federal emergency funds were used for something other than a natural disaster.

DEVELOPING STORY:

A COMMUNITY FIGHTS FOR ITS LIFE

Local citizens banded together to form the Love Canal Home-owners Association, to raise awareness and fight for their cause—and their lives. After increased media attention and exposure on national television, President Jimmy Carter declared a state of emergency for Love Canal. On August 2, 1978, the NYSDOH issued a proclamation recommending that the 99th Street School be closed, that pregnant women and children under the age of two be evacuated, that residents not eat out of their home gardens, and that they spend limited time in their basements. A few days later, New York State agreed to buy and raze 239 homes that were closest to the canal. The EPA decided it would be too dangerous to move the many tons of waste, so the State put up a ten-foot-high fence around a residential area of about seventy acres, defined as the contamination zone.

The homes of another seven hundred families living outside the fenced-in area were deemed safe enough to remain, although they were still measuring high levels of toxins. It was clear that pollutants were extending beyond the fence, affecting the lives of those even farther from the chemical dump site. About a mile away, a fifty-four-year-old woman who had run a hair salon from her basement was now confined to a wheelchair and too frail to work. Her husband had an enlarged spleen (which had to be removed), heart problems, and bone marrow cancer. Their basement sump pump contained a red rubbery sediment—a telltale sign that toxic substances were leeching into their home.

In the spring of 1980, the EPA finished a preliminary study indicating that residents outside the fenced-in contamination zone may have increased chromosomal abnormalities. Blood work on some residents revealed an unusually high number had an elevated risk of seeing cancer, birth defects, or genetic damage in their children.

A President Burned in Effigy and Officials Taken Hostage

The residents-turned-activists demanded that action be taken. Some were so frustrated that they burned figures of President Carter and his wife in effigy. The homeowners told the EPA that if it was so safe in Love Canal, then they could live there, too. Members of the Love Canal Homeowners Association held two EPA officials hostage in their offices, saying that they wouldn't be released until the residents got evacuated. The standoff lasted just five hours, but the homeowners had made their point.

In 1981, President Carter declared his second state of emergency, and the remaining families were relocated. The government bought these additional 550 homes at an average of $35,000 each. About 1,030 families total were evacuated, at a total cost for relocation of about $17 million. Telephone and electrical lines were disconnected. Trees were uprooted. Swimming pools were filled in. Unlike the first 239 homes that were razed, these 550 homes were boarded up, with the intention that one day, the neighborhood could be cleaned up and the houses would be refurbished. Only about 100 people remained behind. Love Canal became America's first toxic ghost town. Rows of houses stood completely empty, many with padlocked doors and broken windows.

Over the next ten years, tens of millions of dollars were poured into cleanup efforts. The seventy-acre contamination zone was capped with a synthetic liner and clay to prevent rainwater from seeping in and carrying more chemical waste outward. The cleanup plan included a tile drain collection system designed to "contain" waste and prevent any outward migration. A trench dug around the canal created an additional barrier. Water collected from the drain system was pumped to an on-site treatment plant.

LASTING IMPACT

Love Canal was a wake-up call for Americans. Many began questioning the effects of other chemical dumping grounds around the country. On December 11, 1980, the US Congress created the Superfund program to clean up the nation's worst hazardous waste sites, and to respond to local and nationally significant environmental emergencies. Since its start, the Superfund has led to the remediation of more than four hundred toxic sites.

In May 1990, the head of the Environmental Protection Agency announced that much of the Love Canal neighborhood was now safe enough for people to live in. By June, more than two hundred families had applied to move back into the area, attracted by bargain prices. They shrugged off fears about lingering chemical poisons. Many who lived in the area considered Niagara Falls as the state's dumping capital; other chemical plants had

been dumping materials in the region for decades, so Love Canal couldn't be much worse than anywhere else. Activists, however, warned that the area was still not safe, and that toxins were still buried underground.

Today, the fenced-off region looks more like a golf course than a poisonous burial ground. Follow-up studies have not established a direct link between the leaking canal and long-term diseases.

In 2011, a work crew repairing a sewer line discovered a pocket of chemical waste about a half-mile from the landfill. In 2013, several families filed suit against Occidental Petroleum, which bought Hooker Chemical. Although they contended that the waste had not been properly contained, state and federal agencies disagreed, saying that the area was continually monitored to make sure pollutants were within acceptable ranges; in fact, a spokesperson from the EPA has called Love Canal the most sampled piece of property on the planet. Nonetheless, some of the new residents have wondered about their health issues, ranging from rashes to miscarriages. ■

Originally, Love Canal was designed to be an ideal working-class community powered with hydroelectricity generated by Niagara Falls. The dream, however, eventually became a nightmare. Wikimedia Commons, photo by Ujjwal Kumar Centralia

THE HEADLINE

LOCATION:
Centralia, Pennsylvania

DATELINE:
1962 – Present

CENTRALIA: THE TOWN THAT NEVER STOPPED BURNING

A community built on coal was then consumed by it

An underground fire that began in a dump more than five decades ago continues to burn today across four hundred subterranean acres of coal seams in the middle of Pennsylvania. The fire spews forth toxic substances such as benzene, hydrogen sulfide, mercury, and arsenic, as well as greenhouse gases like methane and carbon dioxide. Considered the worst mining fire in the United States, the slow-burning inferno has caused a once-thriving town to be almost completely erased from the map, and some scientists estimate that the blaze could continue for more than 250 years.

Old Route 61 had to close after the underground fire caused it to buckle and crack. Today it is covered in graffiti. Carver Mostardi, Alamy Stock Photo

WHAT DO WE KNOW?

For decades, starting in the 1800s, the town of Centralia—about an hour's drive northeast of Harrisburg, Pennsylvania—was a thriving community. For the townspeople, a single product basically fueled their livelihoods: coal, or, more specifically, anthracite. Compared to bituminous coal, which is more common in Pennsylvania, anthracite is much harder to mine. Although it is more difficult to ignite, it burns hotter and cleaner. Underground, coal forms in layers or seams. Anthracite coal seams, however, tend to form at steeper angles, making it more difficult to extract than its bituminous cousin.

Despite challenges in mining anthracite, entrepreneurs figured out how to extract the valuable resource, and large-scale mining operations began in Centralia (originally named Centerville) when the Mine Run Railroad was completed in 1854. By the end of the 1800s, about fourteen mines had opened in the area, and Centralia's population swelled to about 2,700 residents. At that time, the town buzzed with commerce and activity, including five hotels, twenty-seven saloons, seven churches, two theaters, one bank, and a post office, as well as fourteen general and grocery stores, according to Centraliapa.org.

When the United States entered World War I in 1917, coal production decreased as young men left to serve their country. Following the war, strikes by miners further weakened the coal

Smoke rising from the ground in Centralia, Pennsylvania, site of an underground coal seam fire. Wikimedia Commons,

industry. After the big stock market crash of 1929, one of the town's major employers, the Lehigh Valley Coal Company, went out of business, and the coal industry continued its steady decline.

For a brief time after the start of World War II, hopes rose as the demand for coal climbed once again. As production chugged along, the town of Centralia did not want to miss out on any wealth that might be gained. In 1950, the borough acquired the mineral rights to the grounds beneath the community. At this time, the population hovered just below two thousand.

But the riches from coal were not coming back. Consumers were turning to home heating oil, which cost much less. As the big mines closed, strip miners and some illegal operations took over, scrapping out a living, clawing through the earth in grounds still abundant with anthracite.

As mining operations tore through forests and piles of black waste accumulated, Pennsylvania passed a law in 1956 to regulate strip mining and landfills that often led to mine fires.

THE EVIDENCE

O n May 27, 1962, about 1,600 residents of Centralia were getting ready for the annual Memorial Day celebration. The tight-knit community took great pride in the event. It was a chance to show off their hometown. For many residents, their roots ran deep, with family members going back several generations. Even though the coal industry was not what it was, the community was resilient, with about seven hundred to eight hundred buildings still lining the streets. Locals found Centralia an ideal place to work, live, and raise their children. They worshiped on Sundays (at one point there were seven churches), volunteered with the fire department, relaxed at the Legion Hall, looked after their elderly, and helped their neighbors. By all accounts, it was small-town living at its best.

Burning Trash at a Dump Triggers Disaster

Wanting to put the best face on their town for the holiday, Centralians organized a spring cleanup, hauling their trash to an unregulated landfill where they burned it. (Another version of the story says that it was a local company that dumped the trash, leading to the fire.) Burning trash was a common practice in these parts, and although it might have contributed to air pollution, most trash fires simply petered out. This time, however, things did not go as planned. Because the landfill sat on top of an abandoned strip-mining pit, the fire spread from aboveground to seams of coal below, creeping along to reach tons of anthracite in a web of underground tunnels.

The fire was the beginning of the town's descent into oblivion. The product that had been the town's lifeblood would now become its killer.

TIMELINE

1854	Opening of Mine Hill Railroad allows for coal transport
1856	Two major mines open
1890s	Population swells to 2,700
1900–02	Major mining strikes
1917	Anthracite coal mining begins to decline at the start of World War I
1929	Stock market crash forces Lehigh Valley Coal Company to close, leaving thousands out of work
1945	After World War II, cheap fuel oil further reduces demand for anthracite
1950	Town of Centralia acquires mineral rights to coal beneath the borough
1950s–1960s	Strip mining and small coal operations continue
May 27, 1962	The fire begins at a local landfill
Mid-1960s– early 1970s	Twenty-three mining operations go out of business. Fly ash barrier attempted but fails
February 14, 1981	Todd Domboski falls into a hole as ground caves in — population hovers around 1,100
1983	A big portion of Route 61 collapses
1984	Congress allocates $42 million to relocate residents.
1990	Population: 63
1992	Commonwealth of Pennsylvania invokes eminent domain against remaining residents
2002	Centralia Post Office permanently closes
2016	Population: 5

DEVELOPING STORY:

A HOT TIME IN THE OLD TOWN

Centralia did not disappear overnight. The fire crept along steadily over the decades, wreaking havoc as the town fought for its survival. While anthracite can cause flames, it often smolders like a cigarette, giving off deadly carbon monoxide, carbon dioxide, sulfur, and other gases. The fires below could also deplete oxygen levels in enclosed spaces. Firefighters first tried to douse the underground inferno with water and smother it with clay, but smoke kept rising from the ground. Three months after the fire began, officials decided to excavate the area to extinguish it, but as they dug deeper, oxygen only fed the flames.

In a documentary on the fire, resident John Lokitis described how you could see spots of ground glowing red at night, while other areas gave off a blue aura from methane gas. He described how the fire was so close to the surface you could see it burning underground. "It was like looking into a furnace," he said.

As the fire steadily spread below, the trouble affected the people above. Two years after the fire began, twenty-three independent mining operations went out of business, putting about 140 miners out of work, according to the *New York Times*.

By the end of the 1960, residents were experiencing more physical effects from the blaze. One woman told how she received a plant for Easter in April 1969, and just one day after its arrival, it was drooping. By the next day, it was dead. Within a few more days, all the plants in her home had died. Many trees in the area were affected, too—killed off and bleached white from sulfurous fumes.

This same year, William and Janet Birster and their neighbor Marion Laughlin began to get bad headaches. Some other residents were complaining about nausea and unusual bouts of daytime drowsiness. Some kids were getting bronchitis. Some townspeople noticed how matches, lanterns, and candles would sputter out in their homes—a possible sign that there was not enough oxygen.

Miners once used canaries as a warning system to detect poisonous gases such as carbon monoxide and methane. These gases would kill the bird first before it affected the coal miners. The Birsters got a bird for the same reason. On May 18, 1969, Janet Birster woke to find her canary dead in the cage.

Many residents now measured oxygen levels in their home by equipping them with gas detectors. They made sure to stay in rooms where oxygen levels were safer and steered clear of spaces where oxygen was low. When residents on Wood Street sent gas samples from their homes to a lab, they found that oxygen levels were about 2 to 3 percent lower than they should have been. Underground temperatures nearby varied from 65 to 120 degrees, 15 to 70 degrees higher than normal, according to Joan Quigley in her book, *The Day the Earth Caved In*. At a few spots, ground temperatures were reaching 700 to 900 degrees (anthracite burned at 800 degrees). In a 1981 article in *People* magazine, resident Tom Larkin demonstrated how heat from an underground coal fire could fry eggs in a skillet.

A COSTLY PROBLEM

O ne study reported that if all efforts had been made to put out the fire when it first started, the cost would have been about $50,000. From the late 1960s into the early 1970s, however, the US Bureau of Mines invested millions of dollars into another approach to extinguish the fire—pumping fly ash into the ground. Fly ash is a light powdery waste by-product from coal-burning power plants, which had been successfully used in other mine fires to create barriers that blocked them from spreading. But even the thick slurry of fly ash, water, and rocks failed to contain the burning.

Although federal engineers initially deemed the project successful, just a few months later a new hole appeared near the Odd Fellows Cemetery, spewing smoke and steam. In 1975 and 1976, bore holes around town revealed that the carbon monoxide levels were rising. With the fly ash filling deemed a failure, townspeople and government officials were back to the drawing board.

In December 1979, after the mine fires heated eight thousand gallons of gasoline at John Coddington's service station to near boiling point, he gave up the business. This pillar of the community soon became a vacant lot.

In a new effort to improve matters, vent pipes were installed in 1980. Jutting out of the ground, the pipes were surrounded by wire mesh to prevent people from touching them and scalding themselves. According to the website Centraliapa.org, the vent pipes were designed to release carbon monoxide, carbon dioxide, and methane, directing the gases away from homes. At first, as the pipes belched out clouds of noxious fumes, levels of deadly gases inside homes dropped. But the improved conditions didn't last for long, as gas levels in the homes soon rose again. In fact, the fires below seemed to be spreading, and some residents suspected that the pipes were actually feeding the embers with more oxygen.

Doctors warned residents that the gases could aggravate health problems. Nineteen-year-old Catharene Jurgill, who was pregnant with her second child in 1981, was concerned with the well-being of her unborn baby. Although she gave birth to a healthy child, county officials had warned that the gases could pose a health hazard. Catharene's living room was measuring 100 parts per million on a carbon monoxide monitor—about three times higher than the government's threshold. Her kids battled a string of illnesses that did not seem to go away.

A FRUSTRATED TOWN

The townspeople grew increasingly disheartened with the approach of boring holes and testing rather than taking definitive action. Many felt they were getting the runaround from a government bureaucracy. One resident said she did not feel like she was part of America anymore because of the way the government treated them: "The government treats us like cattle and maybe not that good. I wish the officials would take our place for one year and then they'd know what it's like to live like this."

Joan Girolamo was among the embittered residents. At times, her yard could get hotter than her oven at the highest setting, reaching 625 degrees Fahrenheit on the surface. In *People* magazine, Centralian Christine Oakum said that she contacted federal officials to push them to take action. "I wrote a letter to the Secretary of Health. I got a letter back telling me that I shouldn't let my children sniff the cracks in the floor. Can you beat that?"

In a PBS documentary on Centralia from 1982, Robert Belfanti, a state representative in Pennsylvania, said that if the fire were burning near or under Philadelphia, Los Angeles, New York, or Chicago, the response would have been different.

THE EARTH MOVED UNDER THEIR FEET

Not only did the fire produce potentially deadly fumes, it also increasingly disrupted the stability of the earth's surface. More holes in the ground broke through and houses slanted as the soil shifted. The *New York Times* reported how one two-story house in town pitched heavily to one side and looked ready to topple.

In a story on Cracked.com, a Centralia resident said, "Every once in a while, you would come across a deer sticking out vertically with steam billowing out. They looked like they were crawling out. The poor deer had fallen into a sinkhole and had either starved to death or suffocated to death from the fumes." Another resident reported that she had seen a neighborhood cat suddenly disappear forever into a sinkhole. Some townspeople believed that caskets were dropping from their graves, descending deeper into the earth.

The underground mine fire was also taking its toll on the highway into town. The highway buckled and cracked open with fissures in the pavement emitting billowing noxious clouds. In the 1980s, a big portion of Route 61 collapsed and became impassable. Today, about three-quarters of a mile of Route 61 is completely destroyed. Covered with art and writing from visitors, the closed stretch of Route 61 has gained the nickname Graffiti Highway.

FRINGE THEORY

Many Centralians grew to believe that the ongoing fire was part of a plot for coal barons or the government to take full control of the borough and then take advantage of the millions of tons of coal deposits still in the ground. Some estimates pegged the value of the untouched anthracite at around $1 billion. The municipality of Centralia had held the title to the underground resources since 1950. Once the townspeople were driven out, the coal operators would return. In the November 22, 1981, issue of the *New York Times*, Tom Larkin, the then forty-one-year-old president of Concerned Citizens Against the Centralia Mine Fire, supported the theory that Centralia may have been the victim of a conspiracy to propagate the fire deliberately in order to obtain the town's massive coal reserves.

THE TURNING POINT:
WHEN THE GROUND SWALLOWED TODD DOMBOSKI

At Pa. mine-fire site: Afraid to move, afraid to stay

Todd Domboski, who was almost swallowed by a mine-fire hole, stands by fenced-off area in his grandmother's yard.

O n Valentine's Day, 1981, twelve-year-old Todd Domboski and his sixteen-year-old cousin Eric Wolfgang were
working outside when they spotted smoke rising up under some damp leaves in the yard of Todd's grandmother. Think-
ing it might be a fire, Todd walked over to investigate. When he
stopped near the puff of smoke to push aside the leaves, the earth
gave way and he quickly sank up to his knees.

When he struggled to free himself, he plunged farther into the
ground, up to his waist. He put his hands on either side to try and
hoist himself out, but the ground was too soft. In a matter of sec-
onds, he was completely gone from the surface, sinking into the
soft earth as if it were quicksand. Surrounded by the rotten-egg
smell of sulfur, he pushed his heels in front of him to jam his back
against the soil. As he clawed at the earth to slow his descent, his
hand found a tree root and he grabbed hold fast.

Hearing his screams for help, Eric dashed toward the plume
of smoke. He inched toward the hole on his belly and peered in.
About six feet down, he spotted the top of Todd's orange hat. He
yelled for Todd to reach up. With Eric stretching his hand down
and Todd extending his as far as he could upward, they grabbed
hold of each other. Erik was able to yank Todd free, ending his
forty-five-second ordeal.

Without Eric's helping hand, Todd may have slid hundreds of
feet deeper into a hole that could have reached temperatures as

In 1981, twelve-year-old Todd Domboski fell
deep into a smoldering sinkhole. The incident
prompted many residents to push for evacu-
ating Centralia. Newspapers.com

high as 450 degrees. A later analysis found that the temperature just inside the hole was at 160 degrees, and carbon monoxide had reached deadly levels of thirty times the federal government's exposure limits. If Todd had stayed in there much longer, he would have died.

"The smoke was so thick I couldn't see anything," Todd recalled. "I was only in there a minute, but it seemed like an hour." He said that the mud was baked on him so firmly that it couldn't even be cleaned off in a car wash.

When maps of the area were reviewed after his fall, locals discovered that a mineshaft had existed in the yard but had been filled in ages ago. Heat from the continual fires below, however, had loosened the ground, causing the potentially lethal situation.

Todd's plunge into the abyss was a turning point for many in Centralia. About a month later, the town's former mayor collapsed in his apartment above a shuttered gas station and was rushed to the hospital. There was a growing feeling that the town simply wasn't safe, and people had to move.

In the spring of 1981, citizens held a referendum, and in a nonbinding vote of about two to one, said they would demolish their town and move to a new location if that's what it took to put the fire out.

Although at one point James Watt, Interior Secretary under President Ronald Reagan, flatly declared that the mine fire posed "no threat to the health and safety" of the town's residents, he was also suggesting that families be removed from the area after Todd Domboski's fall into the hole. He recommended that the town be demolished, and the grounds and resources be transferred to the state. The Interior Department authorized a $1 million buyout, pricing about twenty-eight homes at $15,000 each, and offering $15,000 in relocation funds per family.

A Town Is Torn Down

The majority of the residents began leaving in 1984, and Congress provided $42 million in funding for their relocation. Many families moved to nearby communities such as Mount Carmel and Ashland. As the government bought up homes and people moved out, their houses would be torn down, completely demolished, leaving no visible sign that anyone had ever been there. Sometimes they'd wait until a whole block was evacuated, and then level the entire block in one fell swoop.

The erasing of Centralia was hastened in 1992 when Governor Bob Casey invoked eminent domain on all properties in the borough. With every building condemned, about five hundred were obliterated. Still, some citizens fought back, initiating a lawsuit and arguing that they had a right to stay.

After much legal wrangling, seven property owners finally settled with the state in 2013, coming to terms that allowed the current owners to have the right to remain in their properties until they died, although these properties were now officially owned by the State.

The Assumption of the Blessed Virgin Mary Ukrainian Greek-Catholic Church on North Paxton Street in Centralia overlooks the abandoned town. Photo by Simon Burchell

LASTING IMPACT:
COAL FIRE'S
DESTRUCTIVE POWER

In the early 1980s, the population of Centralia stood at about one thousand residents. By 1990, it had dropped to sixty-three, and then by 2010, only ten remained. In the fall of 2016, Kathi Womer moved out of Centralia, bringing the count to just five. What's left of Centralia is not even a ghost town; it's really almost no town at all. The government's plan to evacuate the borough basically wiped out the entire identity of a community and its people. By 2002, the local post office had been torn down, and the population had dwindled to such a degree that the US Postal System took away Centralia's zip code. Without a zip code, mail delivery stopped. Remaining residents had to establish post office boxes in neighboring towns. There is no local police force. All the highway signs have been replaced, removing "Centralia." The name was removed from the city municipal building (one of the town's last remaining structures), and the name no longer appears on maps.

Tourists wander through, curious about the town that disappeared, and poking around the smoldering patches that indicate that a fire still burns below. The few remaining citizens say that the tourists are more destructive than the fire. A few visitors have even chipped off pieces of homes for souvenirs. Many leave behind garbage. The Assumption of the Blessed Virgin Mary Church, a Ukrainian Greek-Catholic church, located on North Paxton Street, overlooks the remnants of Centralia. The cemetery remains as well— spared destruction because it lies just outside the mine fire impact zone. Some former residents say they still want to be buried there—that it's the only way they can return to their hometown. In the meantime, the winds of change continue; today, wind turbines line the mountain ridges surrounding Centralia.. ■

COAL MINE FIRES: A WORLDWIDE PROBLEM

According to the Office of Surface Mining and Reclamation and Enforcement, more than two hundred coal fires are burning above- and belowground in the United States, in fourteen different states. Coal fires are not unusual—they rage on every continent except Antarctica. Although the Centralia fire is expected to continue for decades, it is not the longest-burning coal fire of all time. That title goes to Burning Mountain in Australia, which has been ablaze for six thousand years. No one is certain how the coal deposits caught flame, but some theories suggest a lightning strike, a forest fire, or aboriginal activities are to blame.

Coal fires are one of the largest sources of carbon dioxide pollution on Earth. An article in *Earth: The Science Behind the Headlines* reported that China loses two hundred million tons of coal each year to mine fires, according to the International Institute for Geo-Information Science and Earth Observation, based in Enschede, Netherlands. Burning coal not only increases carbon dioxide and carbon monoxide levels; it also produces harmful toxins such as arsenic, selenium, fluorine, sulfur, lead, copper, bismuth, tin, germanium, and mercury.

Burning Mountain in Australia is the longest-burning coal fire in history. At the summit, the dirt has changed to a bright red because of oxidation of iron compounds in the soils. Wikimedia Commons, photo by Beruthiel

THE HEADLINE

GILMAN: A DEADLY WATER SUPPLY DRIVES A TOWN TO EXTINCTION

LOCATION:
Gilman, Colorado

DATELINE:
1886–1985

Tons of mining waste made Gilman into a modern-day ghost town

In November 1984, the popular luxury ski resort in Beaver Creek, Colorado, was in full snowmaking mode as ski season was just starting. As the workers ran snowmakers to freshen the slopes, they noticed a strange orange tint coloring the white landscape. The snowmakers drew their water from miles away in Eagle River. Toxic pools from the Eagle Mine had overflowed, sending a bright orange stream of toxic sludge surging into the rushing waters.

For decades, the Eagle Mine had been the economic engine for the town of Gilman, about fifteen miles away from Beaver Creek. Perched on a hillside of Battle Mountain, hundreds of feet above the Eagle River, Gilman residents depended on the mining industry for their livelihoods. Now, the small town on US Highway 24, a two-hour drive west from Denver, was completely abandoned, but its legacy lived on as deadly contaminants from the mine polluted the region, and efforts to clean up were just beginning.

TIMELINE

1878	First mine in Gilman area opens
1886	Gilman established as a town
1912	New Jersey Zinc buys mines in Gilman
1950s–1960s	Eagle Mine hits peak production and Gilman thrives
1983	Colorado files a lawsuit against the mine's owners and former owners for environmental damages
1985	The town of Gilman is abandoned
1986	The Environmental Protection Agency declares Gilman a priority Superfund site; intensive cleanup efforts begin
1990	A water treatment plant begins operations
2001	Gilman is declared environmentally safe

Poisons in the ground and water from local mining reached such toxic levels that residents of Gilman had to leave their homes overnight. Wikimedia Commons, photo by Matthew Trump, 2005

WHAT DO WE KNOW?

Mining in and around Gilman dated back to 1878, when the Little Ollie Mine began operations at the beginning of the Colorado silver boom. On May 5, 1879, a judge by the name of D. D. Belden unearthed what would become Belden Mine, which tapped into a vast lode of silver. That same year, the Iron Mask Mine struck pay dirt, hitting rich reserves of zinc and lead. In 1882, when the Denver and Rio Grande Railroad extended tracks to the base of the cliff below the mines, transporting valuable minerals became much easier and more profitable.

As the mining industry grew in 1886, prospector and judge, John Clinton, from nearby Red Cliff, founded the town that would become Gilman. Originally, the community was named Clinton, but when it was discovered that another town was already named Clinton, the locale changed its name to Gilman, in honor of Henry Gilman, a well-liked superintendent of the Iron Mask Mine.

The Groundhog Mine became one of the most prosperous operations in Gilman, producing gold and silver ore into the 1920s. Other mines included the Ida May, Little Duke, May Queen, Kingfisher, Little Chief, and Crown Point. By 1899, the town had grown to about three hundred—big enough to support a school, a boardinghouse, and its own local newspaper, the *Gilman Enterprise*.

In 1912, when New Jersey Zinc began buying up all the mines, the Iron Mask changed its name to the Eagle. In general, mining operations in the area shifted away from silver and gold production and more toward zinc and lead. Although zinc prices plunged during the Great Depression, New Jersey Zinc weathered the economic hard times, raking in huge profits a decade later when demand once again soared. For Eagle County, the prosperous mine provided a substantial property tax revenue. The company hit a peak in 1951, generating $12 million in product. At its high point, the mine employed about seven hundred people, as the Eagle became the world's largest zinc mine.

New Jersey Zinc opened a post office and built a clubhouse for its workers. In 1960, the population hit an all-time high. Workers could rent homes for as little as $46 per month. In many ways it was an idyllic small-town community. Kids bought penny candy from Murphy's Store. Families bowled at the bowling alley in the company clubhouse. Residents fished in the nearby creeks.

In January 2018, OutThereColorado.com published an interview with former Gilman resident Bill Squires, who worked at the mines during their heyday. He described decent pay, the good union work, homes made of hardwood, yards blooming with gardens, and happy, simple times in the clubhouse.

The good times, however, could not last forever. The industry steadily declined in the 1970s and into the early 1980s, when prices hit rock bottom.

THE EVIDENCE

In 1966, Gulf and Western acquired New Jersey Zinc and ran the business until 1977, when it halted most of the mining operations. In 1983 the State of Colorado filed a lawsuit against the mine's owners and former owner for environmental damages resulting from the mine operations.

The ground had been severely polluted by heavy metal contamination from about eight million tons of mining waste deposited into the ecosystem. High levels of arsenic, cadmium, copper, lead, and zinc had polluted the soils, structures, surface water, sediments, and groundwater across the site. All the metals found at Gilman could have harmful health effects in humans if ingested. For example, arsenic ingested through food, water, or the air can cause everything from a sore throat to death.

The contaminants reduced fish populations and other aquatic life in the Eagle River downstream from the mine and its waste piles. The contaminants also posed a threat to two drinking water wells used by the town of Minturn, six miles away. Airborne particles were also a concern, and contamination was destroying the nearby Maloit Park wetlands area as well.

In an interview with the *New York Times* in 2005, Caroline Bradford, a longtime resident who worked as executive director of the Eagle River Watershed Council, a nonprofit group working to revive the river, said, "Everything was dead. No fish, no bugs, nothing."

Colorado has always been a popular destination for skiers. Personnel at Beaver Creek ski resort knew something was wrong when their snowmakers started spreading orange snow. Flickr, photo by Felix Abraham

DEVELOPING STORY:
A TOWN DISAPPEARS OVERNIGHT

By 1984, the town had been sold to the Battle Mountain Corp., which evicted all the residents the following year. With the electricity completely shut down, all properties were quickly evacuated. One resident who grew up there wrote that Gilman became a ghost town almost overnight. "We had little notice, and what we couldn't fit in pickup trucks or didn't need stayed behind."

Most of the buildings were left standing, their contents intact. The hospital is shuttered, with X-rays scattered about. Office equipment and file cabinets gather dust in various buildings. Empty homes line the hillside, some tagged with graffiti, walls cracked and chimneys crumbling. The rusty swing set remains in the schoolyard. Vehicles remain unmoved, including a '57 Chevy dump truck and a '65 Chevy station wagon.

With no one to tend the mine, the tunnels began to flood. Downstream from the mine, the Eagle River took on an orange tint. In 1986, the Environmental Protection Agency declared 235 acres of Gilman a Superfund hazardous waste site, and placed the town on their priority list. The Agency posted signs to keep trespassers off the property: HIDDEN AND VISIBLE DANGERS AND RISK OF INJURY OR DEATH.

Over the next decade and a half, tons of mining waste was relocated. From 1986 through 1990, six openings to the Eagle Mine workings were permanently plugged to prevent the direct flow of mine water into the Eagle River. The EPA set up a water treatment plant in 1990 and continues to monitor water quality today.

Gilman was once a bustling blue-collar community. Newspapers.com

Residents of Gilman left so quickly that they left many of their homes and offices filled with furniture, supplies, and other belongings. Photo by Jonathan Haeber, reprinted with permission

LASTING IMPACT

In 2001, after extensive cleanup efforts, the EPA declared that public health risks had been significantly reduced and the area was once again safe. Studies show that the fish and insect populations in the river have rebounded. A broad coalition of citizens, local conservation organizations, and municipal and county government continues to ensure the environmental safety of the region. As of 2017, total cost for cleaning up the mining pollutants was estimated at about $60 million.

The Eagle Mine Water Treatment Plant treats an average of 221 gallons per minute, or 116 million gallons per year. During the process, the plant removes 178 pounds of metals per day. Although trout have returned, residents have not come back to Gilman. Some who left the mining industry found opportunities in the growing ski resort business. One day the town may be reborn. In 2004, developers began initial plans to transform lands around Gilman, including portions of the Eagle Mine Superfund site, into a ski resort and possible residential use. ∎

The old Eagle Mine in Gilman was the source of many deadly contaminants. Flickr, photo by El-toro

PART V

FADING ICONIC AMERICA: LOST HIGHWAYS, DRIVE-INS, AND DINERS.

The Jackson Sun · Thursday, May 25, 2017 · 5B

LIFESTYLES

Expiring law could leave Route 66 towns without funding

RUSSELL CONTRERAS
ASSOCIATED PRESS

ALBUQUERQUE, N.M. – Route 66, the historic American roadway that linked Chicago to the West Coast, soon may be dropped from a National Park Service preservation program, which would end years of efforts aimed at reviving old tourist spots in struggling towns.

A federal law authorizing the Route 66 Corridor Preservation Program is set to expire in two years, and some lawmakers are working to save the program or get Congress to designate Route 66 as a National Historic Trail. That designation would set aside preservation funds annually.

The deadline, first reported by The Herald-News in Joliet, Illinois, also has Route 66 enthusiasts and preservation advocates scrambling to make sure the program or an alternative is maintained for the "Mother Road."

The program is credited with helping bring back to life forgotten landmarks along the route, many in disrepair because of sharply lower Route 66 traffic. Development of the interstate highway system after World War II diverted motorists away from Route 66 and economically hurt communities along the road.

A bipartisan bill in Congress to designate Route 66 as a National Historic Trail, sponsored by Rep. Darin LaHood, an Illinois Republican, is supported by 12 other members of Congress from Illinois, Kansas, Oklahoma and California.

New Mexico has the longest stretch of Route 66 passing through various communities, but state Tourism Secretary Rebecca Latham said she did not

know whether the program would have an adverse impact on the state if it is not eventually refunded.

Preservationists fear that small towns along Route 66's 2,500-mile path will miss out in much-needed investment if the funding program is not extended or if the route does not get the historic trail designation, said Frank Butterfield, director of the nonprofit group Landmarks Illinois.

"Route 66 runs through a lot of very small towns where there is not a lot of economic development," Butterfield said. "It's been quite impactful so

program manager for the Route 66 Corridor Preservation Program in Santa Fe, New Mexico.

The program has helped fund projects like the El Vado Motel neon sign restoration in Albuquerque, New Mexico, and the Baxter Springs Independent Oil and Gas Station restoration in Baxter Springs, Kansas.

Other grants are being used to repair the roof of the Historic Navajo County Courthouse in Holbrook, Arizona, and restore a free-standing neon sign of the Donut Drive-In in St. Louis. Some places that have received preservation funding are listed on the National Register of Historic Places. Tourism and local officials believe the grants help revitalize forgotten structure and spur tourism, especially by international visitors who flock to Route 66.

Decommissioned as a U.S. highway in 1985, Route 66 went through eight states, connecting two with friendly diners in welcoming small towns.

It was once an economic driver for small towns from Illinois to California. Nat King Cole famously sang "Get Your Kicks (on) Route 66" in a 1946 hit that has been remade by countless other groups.

Use of Route 66 dropped significantly after highways were built as part of the interstate system, forcing businesses to close and leaving others

it would be a great loss to the towns where Route 66 passes."

The uncertainty comes as the Trump Administration is proposing deep cuts to domestic spending and various agencies.

Established in 2001 by Congress, the Route 66 Corridor Preservation Program began as an effort to save the historic trail designation, said

in disrepair.

Barthali said the program has funded 20 projects in New Mexico and given out $340,000 in grants, including $8,000 to help fix up the windows and neon sign of the historic Blue Swallow Motel in Tucumcari, New Mexico. That grant helped attract an additional $27,000 for restoration.

A car heads toward Albuquerque, N.M., on historic Route 66 in 2014. A federal law authorizing the Route 66 Corridor Preservation Program is set to expire in two years, and with it would go millions of dollars in grants for reviving old tourist spots in struggling towns.

ROAD TRIP

Get nostalgic kicks on old Route 66

By John Stanley
THE ARIZONA REPUBLIC

Motorcyclists headline trip ride on historic Route 66 after stopping for lunch at Delgadillo's Snow Cap Drive-In in Seligman.

Getting there: About 490 miles, round trip, from central Phoenix. Take Interstate 17 north about 145 miles to Flagstaff, then Interstate 40 west about 73 miles to Seligman, via Exit 123. In Seligman, head west on Route 66 about 25 miles to Grand Canyon Caverns, then continue about 10 miles to Peach Springs.

Don't miss: Diamond Creek Road is the only place between Lees Ferry and Lake Mead where you can drive down to the Colorado River in the Grand Canyon. Start in Peach Springs and pick up a permit ($15 per person, plus tax) at Hualapai Lodge, between mile markers 103 and 104 on the southern side of Route 66. Only the first mile of the 40-mile round trip is paved, so plan on a drive of two hours. A high-clearance four-wheel-drive vehicle is recommended. The road may be closed after storms.

E2 · Pantagraph · Sunday, June 4, 2006

GO!FESTIVALS

www.pantagraph.com

Tour celebrates 'Mother Road'

By Scott Richardson

BLOOMINGTON — Time to get your kicks on Route 66 when the 17th annual Route 66 Association of Illinois 2006 Motor Tour celebrates the allure of the "Mother Road" Friday through June 11.

Once the main east-west artery from Chicago to Los Angeles, the highway was designated Route 66 in 1926. Though decommissioned in 1984, the road continues to draw visitors from around the globe to savor what remains and what's been restored of the mom and pop restaurants, gasoline stations and motels that gave the highway its unique character. It's been called a "2,400-mile-long museum of America."

ROADSIDE ATTRACTION

Right: Sites like this gasoline station in Odell have been restored to their appearance during the heyday of Route 66. Signs on Route 66 designate tourist spots like the Dixie Truckers Home in McLean as Route 66 attractions. The truck stop was once home to the museum.

The Tom Teague Ambassador's Award goes to Marilyn and Durelle Pritchard of Pontiac. The late Tom Teague served as executive director of the Route 66 Museum.

The annual motor tour starts with the Mother Road Festival in Edwardsville. Breakfast in Springfield and stop in Lincoln Saturday night. The final day starts with breakfast in Atlanta and ends with a rain-or-shine car show

and sock hop in Pontiac.

Many members drive classic cars, but they are not required. Join or leave the motor tour at any point.

Tour drivers can stop at several sites along the way, including Odell, where the association has restored a gasoline station and Route 66 Hall of Fame Museum.

THE
HEADLINE

ROUTE 66: AMERICA'S ROAD TO NOWHERE

LOCATION:
Chicago to Santa Monica

DATELINE:
1926–1985

When the famous highway was decommissioned, it meant the end of the road for many small towns

No other thoroughfare in America captures the country's love affair with the open road more than Route 66. At its height, the first fully paved highway in America meandered through small-town USA, stretching 2,448 miles from Chicago to Santa Monica. Along the way, it wound through Missouri, Kansas, Oklahoma, Texas, New Mexico, and Arizona. It's no wonder the highway soon became known as the Main Street of America.

When World War II ended, families enjoyed a new prosperity. During their leisure time, many wanted to explore their own great country. With car culture booming, Americans filled up their tanks, piled in their autos, and took off on vacation along Route 66. Unlike most of the relatively straight highways in the country, this one twisted and turned through the heartland. The highway came to represent the freedom of the open road at a time when Americans were enjoying the postwar prosperity of the 1950s.

Route 66 still stirs nostalgia for diners, middle-of-nowhere truck stops, Chevys, convertibles, and James Dean. For decades, small communities thrived along the roadside, as it served as a major artery for commerce and tourism. In time, however, bigger roadways with more lanes and more-direct routes made the old two-lane charmer obsolete. While portions of the highway remain, many of the communities that once prospered along Route 66—like Glenrio, Texas—have declined or turned into ghost towns.

The famous Gemini Giant continues to greet customers outside the Launching Pad Cafe on Route 66 in Wilmington, Illinois. Wikimedia Commons, photographer unknown

WHAT DO WE KNOW?

I n the 1920s, automobile sales accelerated at a brisk clip as the cost of owning a car became more affordable. Automobile registrations, which numbered eight million in 1920, almost tripled to twenty-three million by the end of the decade. As cars rapidly rolled off the assembly line, the country needed more and better roads. The American Association of State Highway Officials recommended a system of major roadways that resulted in twelve odd-numbered interstate routes running north to south and ten even-numbered ones running east to west. Two businessmen, Cyrus Avery of Tulsa, Oklahoma, and John Woodruff of Springfield, Missouri, spearheaded the movement to build a highway connecting the Midwest to the West. Because of their efforts lobbying the federal government, the Federal Aid Highway Act of 1925 funded the construction of Route 66.

As envisioned by Woodruff and Avery, this major road was designed to wind through the modest towns of America so these communities could prosper from a major thoroughfare. The first third opened on November 11, 1926, with the entire route reaching completion twelve years later. The highway project was a godsend for many families, as it employed thousands of able-bodied men during the Great Depression.

Swooping south and west from Illinois to California, the road traversed mostly flatlands, making it popular for truckers. Oklahoma became most associated with the route, with the longest

Just over three miles from Route 66, Ed Galloway's Totem Pole Park in northeastern Oklahoma features the "World's Largest Concrete Totem Pole." Wikimedia Commons, photo by Dustin M. Ramsey, 2004

The Snow Cap Drive-In along old Route 66 in Seligman, Arizona, has been attracting hungry travelers for decades with funny signs such as "Sorry, We're Open!" Flick, photo by inkknife

TIMELINE

1925   The Federal Aid Highway Act provides funding for Route 66

1926   The first portion of Route 66 opens

1938   Route 66 completed; becomes first continuously paved highway

1946   Nat King Cole releases hit single, "Route 66"

1956   Federal Aid Highway Act paves way for interstate system that makes Route 66 obsolete

1960s   TV series "Route 66 " hits the airwaves

1985   Route 66 decommissioned.

1999   Federal government provides $10 million to help preserve surviving portions of Route 66

segment of about four hundred miles. The thoroughfare gained the nickname "the Will Rogers Highway," honoring the famous humorist and Oklahoma native son.

During the Great Depression in the 1930s, severe dust storms and drought ravaged the Southern Plains area from Texas to Nebraska. The harsh conditions forced hundreds of thousands of people from the region (often referred to as "Okies") to migrate westward. Many families packed up all their worldly possessions and headed toward California along this "road to opportunity." John Steinbeck, who immortalized the exodus in his novel *The Grapes of Wrath*, wrote "66 is the mother road, the road of flight."

During World War II, the federal government designated Route 66 as a strategic defense highway. After the war, many returning soldiers followed the same passageway, looking to settle in the warmer climes of the Southwest. The romance of this road grew even greater in 1946 when Nat King Cole released the single "Route 66," written by Bobby Troup. The whole country sang along about getting their "kicks on Route 66"—a phrase that helped to secure the road's place in the history of pop culture. In the 1950s, those who identified with the "Beat Generation" embraced the road. This movement of young writers and artists, including Jack Kerouac and Allen Ginsberg, rejected the conventions of society and valued free expression. The long highway called to them as a way toward "finding" themselves.

In the early 1960s, a popular TV series called *Route 66* told the story of two adventure seekers traveling from town to town in their Corvette. Roadways like this contributed to the rise of fast food, motor inns, billboards, and drive-up businesses, along with the concept of picnic areas and rest stops.

THE EVIDENCE: BIGGER HIGHWAYS TAKE OVER

ROUTE 66

In some ways, when President Dwight Eisenhower signed the Federal Aid Highway Act in 1956, it was the beginning of the end for Route 66. This law provided funding to connect America through bigger and better superhighways. By the 1970s, portions of Route 66 were widened to four lanes, and a series of interstates replaced most of the original road. Many of these new roadways paralleled the original thoroughfare. By 1977, Interstate 55 took the place of Route 66 in Illinois, and many of the "66" road signs were removed. The entire road was decommissioned on June 27, 1985, although travelers can still find sections marked as "Historic Route 66."

DEVELOPING STORY:
PRESERVING AN AMERICAN ICON

LEFT: **Cars like this 1931 Studebaker once filled Route 66 as America fell in love with the automobile and hit the road.** Wikimedia Commons, photo by Finetooth

RIGHT: **The Highway Diner in Winslow, Arizona.** Library of Congress

BELOW: **The Route 66 Motel in Needles, California.** Wikimedia Commons, photo by Renjishino

Travelers can still stay overnight in a wigwam at Wigwam Village in Holbrook, Arizona. Library of Congress

Since 1985, supporters have formed groups to help preserve what's left of the famous highway, having the spots designated as National Historic Places or National Scenic Byways. In 1999, President Bill Clinton signed a bill allotting $10 million to help maintain the remnants of Route 66.

Although about 85 percent of the road is still drivable (even if it's unmarked), its decline is clearly evident. The big towns along the route are still there, such as Springfield, Illinois; St. Louis and Joplin, Missouri; Tulsa and Oklahoma City, Oklahoma; Amarillo, Texas; Albuquerque, New Mexico; Holbrook and Flagstaff, Arizona; and Needles, Barstow, and San Bernardino, California. But when the Main Street of America died, it took several towns with it. Some today call Route 66 the Avenue of Broken Dreams.

Portions of the 1940 film The Grapes of Wrath were filmed on Route 66 in Glenrio, Texas.
Photofest

ONE TOWN'S TALE

Glenrio, which straddles the borders of Texas and New Mexico, started out with great promise in 1903. After the Chicago, Rock Island, and Pacific Railroad came through the area around that time, a growing number of small farmers settled in the region, and the community soon had a few grocery stores, a hardware store, service stations, and coffee shops. The town continued to pick up steam when Route 66 came through in the 1930s. The destination proved to be a favored stopping point for travelers, and motels, diners, and gas stations did a solid business catering to them. A "welcome station" was opened for those crossing the state borders. Shortly after the highway opened, John Ford arrived in 1938 to shoot some scenes for the movie version of *The Grapes of Wrath*.

Although the town thrived through the 1940s and into the early 1950s, it suffered a double whammy of bad fortune that led to its demise when the local railroad depot closed in 1955 and Interstate 40 opened, diverting traffic from the small community and sealing its death warrant. For years, the Ehresman family had made a living in Glenrio running a tourist shop, gas station, and grocery store. As business dried up in the late 1950s, the family pulled up its roots and headed five miles west to Endee, New Mexico, but it wasn't long before Endee, too, lost business because of the interstate. Both communities are now ghost towns, with a few shuttered buildings still standing, like the Little Juarez Café and the State Line Motel, with its sign that reads "First in Texas" or "Last in Texas," depending on whether you're heading east or west.

Glenrio and Endee were among the dozens of towns that faded away along the old highway, including Newkirk and Dilia, New Mexico; Erick and Afton, Oklahoma; and Oatman, Arizona. For miles heading west out of Springfield, Missouri, local economies collapsed in Plano, Halltown, Heatonville, and Carthage.

Americans maintain a nostalgic love for Route 66 today. Wikimedia Commons, photo by Jpatokal

Young people continue to seek the romance of the open road along Route 66. Wikimedia Commons, photo by Christopher Michel

La Mesa Motel. Wikimedia Commons, photo by Shoshone

LASTING IMPACT:

NOSTALGIA KEEPS ROUTE 66 ALIVE

There are signs of life returning today. Although Paris Springs Junction, Missouri, is almost completely deserted, Gary and Lee Turner bought the old Sinclair filling station in the late 1970s, re-creating its original mid-twentieth-century ambience and making it a new tourist destination.

In addition to this revamped filling station, a few other quirky motels and roadside attractions still hang on, surviving on those tourists who are eager to tap into nostalgia for the America that was. The thirty-foot-tall Gemini Giant, wearing his space helmet and clutching a rocket, still greets those coming to Wilmington,

Illinois, standing watch over the entranceway of the Launching Pad drive-in restaurant.

Just three miles off the Mother Road in Foyil, Oklahoma, Ed Galloway created his own tribute to Native American culture when he built Totem Pole Park between 1937 and 1948. Tourists can still come here to see the world's largest concrete totem pole made from twenty-eight tons of cement, six tons of steel, and one hundred tons of sand and rock. Established in the 1940s, the Jack Rabbit Trading Post near Joseph City, Arizona, continues to sell curios, and visitors can ride the large fiberglass rabbit sitting out front. The Snow Cap Drive-In, opened in 1953, attracts hungry throngs with a menu offering "Cheeseburger with Cheese" and "Dead Chicken," as well as a neon sign shining "Sorry, We're Open."

Road-weary travelers can go back in time when they check in at the Wigwam Village Motel #6 in Holbrook, Arizona, or the Wigwam Motel in San Bernardino, California. Both motels feature individual tepees furnished with beds and equipped with a toilet, sink, and shower. In Amboy, California, Roy's stands out in the Mojave Desert as a 1950s-style gas station and cafe. The tiny town in the middle of nowhere supported its own school, church, and airport as well. Sightseers also come to check out the nearby six-thousand-year-old volcanic cinder cone called the Amboy Crater. Route 66 passes by or near many other natural wonders, including the Painted Desert, the Petrified Forest, and the Grand Canyon.

About an hour west, the Bagdad Café serves hungry customers. Originally, the restaurant, which gained famed in the eponymous 1987 German movie, operated in the town of Bagdad, California, which ceased to exist after Interstate 40 opened in 1973. Although named after one of the largest cities in the Arab world, Bagdad (spelled without the "h" in this case) was completely razed and has become a ghost town, with only a few foundations and one solitary tree marking its location. The Sidewinder Cafe in nearby Newberry Springs has capitalized on the enduring popularity of the film, changing its name to Bagdad Cafe in 1995.

The Eagles captured some of the romance of Route 66 in their song "Take It Easy," which includes the lyric "Standing on a corner in Winslow, Arizona." Wikimedia Commons, photo by Marine

Dustin Hoffman and Tom Cruise travel along the Mother Road in the hit movie *Rain Man*. Photofest

NEWER ATTRACTIONS

Some newer sites along Route 66 today still capture the spirit of the road. In Amarillo, Texas, the public art installation known as Cadillac Ranch features ten old Cadillacs buried nose-down in the plain, showing off a variety of tailfins. The display, which opened in 1974, serves as a monument to the Golden Age of the American automobile. In 1999, The Eagles' song "Take It Easy" was commemorated by a statue of a man holding a guitar on a corner in Winslow, Arizona. In 2000, Elmer Long started an enchanted folk-art forest known as the Bottle Tree Ranch in Oro Grande, California. Here, thousands of green, blue, brown, and clear bottles hang on mostly metal "trees," clinking in the wind.

If you're traveling Route 66 from end to end, a good beginning point may be Lou Mitchell's in Chicago, often called "the First Stop on the Mother Road." The diner has been serving hearty breakfasts and lunches since 1923. At its western terminus, Santa Monica provides a celebratory endpoint for any trek along the highway. The "Pleasure Pier" entertains all comers with various rides and amusements, including a Hippodrome building built in a California-Byzantine-Moorish style, and a 1922 carousel complete with hand-carved horses. Plus, the highway ends right at the Pacific Ocean, providing a spectacular seaside view. ∎

The **Blue Whale of Catoosa in Oklahoma has become one of the most recognizable attractions on old Route 66.** Photo by Rich Lee, used with permission

ROUTE 66 LIVES ON IN POP CULTURE

Besides being made famous in a hit song and a TV series, Route 66 and its attractions have provided fodder for other creative works:

◆ Bruce Springsteen's "Cadillac Ranch" was inspired by the roadside installation with the same name.

◆ Many scenes in the movie *Rain Man*, starring Dustin Hoffman and Tom Cruise, were shot along the roadway, including scenes at the Big 8 Motel in El Reno, Oklahoma, which later became the nondescript Deluxe Inn.

◆ A minor league baseball team in San Bernardino, California—the Inland Empire 66ers—took their name from the famous route.

◆ Sega produced a video game called *The King of Route 66* that features such iconic stops as the Bagdad Cafe and the Santa Monica Pier.

TWO OTHER LOST HIGHWAYS

Americans tend to love exploration and adventure. As the country grew, so did its passageways for travel. But as with Route 66, many of the nation's famous byways went to the wayside. Once-bustling arteries, clogged with travelers and commerce, have either vanished or transformed into quieter tourist attractions and historical sites.

The Oregon Trail

For American pioneers heading west in search of new opportunities, the Oregon Trail was the "superhighway" of its time, providing families in their covered wagons, it with a 2,200-mile route from Independence, Missouri, in the middle of the country to Oregon City, just south of Portland, in the fertile Willamette Valley. Fur trappers and traders began forging the trail in the early 1800s, followed by missionaries and settlers. In 1836, the first wagon train made it out to Fort Hall, Idaho. Those heading farther west would often ditch their wagons in Idaho as they continued on into rougher terrain.

In 1843, Dr. Marcus Whitman guided a wagon train from Idaho to Walla Walla, Washington, initiating what was known as "the Great Migration." This territorial expansion in the 1840s was called "Manifest Destiny"—a belief that God intended the United States to control most of North America. Between 1843 and the Civil War, about four hundred thousand American pioneers trekked westward along the trail. The journey was not easy. River crossings posed one of the greatest dangers. Rushing waters took the lives of hundreds. Ferry businesses charged outrageous amounts to take families and wagons over treacherous currents. The trail took its toll on feet as well; many walked the entire distance. Accidents (such as falling under a wagon wheel), severe weather, and disease also wiped out many settlers. The fatal bacterial disease cholera claimed thousands of lives, along with malaria and yellow fever. Cholera was a quick killer; the infected could go from healthy to gravely ill in a matter of hours.

Although old Western movies depict Native Americans as a constant threat to pioneers, hostile actions between the settlers and Indians were rare. The first part of the trail cut through territory where the Pawnee tribe resided to the south and the Cheyenne, to the north. Records show that many of the interactions were friendly. The newcomers and the Natives often traded goods. In some cases, the Indians helped to round up stray cattle, rescued emigrants who fell into rivers, or dislodged wagons that had become stuck.

Some stretches of the old Oregon Trail are just as they were when pioneers traveled them in wagons. Bureau of Land Management, Oregon National Historic Trail

While the trail reached all the way to Oregon, most veered off the path before the final stop, settling in Wyoming, Idaho, or Utah, or making it as far as California. The last wagon trains on the Oregon Trail disappeared in the 1880s, gradually declining after the completion of the Transcontinental Railroad in 1869. With this transcontinental link, people could now travel coast to coast by train more quickly and safely. The route largely followed the already-established Oregon, Mormon, and California Trails.

Tourists can still check out remnants of the Oregon Trail today, including the deep grooves dug by wagon wheels in Guernsey, Wyoming; the trail ruts that cut through the countryside at Rock Creek Station, Nebraska; or the names carved by pioneers in Independence Rock in Casper, Wyoming.

The Lincoln Highway

Predating Route 66 by more than a decade, the Lincoln Highway, which was completed in 1913, provided America with its first cross-country automobile road, extending coast to coast from Times Square in New York City to Lincoln Park in San Francisco. Originally measuring a total of 3,389 miles, the Lincoln Highway had some paved portions patched together with an assortment of dirt, stone, and gravel pathways to complete one long thoroughfare. According to an article posted on the Federal Highway Administration website, a trip from the Atlantic to the Pacific in 1916 was still a long journey, taking twenty to thirty days.

The artery included historic roads such as a thoroughfare in New Jersey created by the Dutch colonists before 1675; a military trail in Pennsylvania built by the British in 1758, when fighting the French and Indian War; parts of the Mormon Trail; a portion of the Pony Express Route; and a section of Ridge Road, used by Native Americans in Ohio for centuries. Lengths of the highway, called "seedling miles," were created to demonstrate the superiority of cement-covered roads.

After the US government established the Numbered Highway System in 1926, most of this roadway became Route 30. Although Lincoln Highway has vanished, reminders exist today. Some stretches of road in Pennsylvania and Iowa are still named Lincoln Way or Lincoln Highway. In 1992, some fans of the roadway established the Lincoln Highway Association, to "preserve, and improve access to, the remaining portions of the Lincoln Highway and its associated historic sites."

OTHER LOST ICONS

Disappearing Drive-Ins

June 6, 1933, was a big day for American auto drivers and movie fans. On this date in history, cars packed into the lot at Park-In Theaters in Camden, to be the first to watch a movie without leaving their cars. Although the film shown that night, the comedy *Wife Beware*, had been released three years earlier, the crowds didn't mind. Eager for a new experience, they shelled out the twenty-five cents per car and twenty-five cents per person.

Auto parts salesman Richard Hollingshead came up with the idea after experimenting with a Kodak projector placed on his car hood, showing the movie on a screen he'd hung in some trees. For his drive-in theater, he designed a series of ramps that placed cars in elevated semicircles for optimal viewing of the screen. Speakers were centrally located to broadcast the sound to a lot full of cars.

In 1947, a group of six teenage girls picketed the Aurora Drive-In in Seattle, complaining that outdoor movies were stealing jobs from babysitters. Wikimedia Commons, Seattle Municipal Archives

The 66 Drive-In theater on the old Route 66 in Carthage, Missouri, still shows first-run films. Wikimedia Commons, photo by Jackbrown

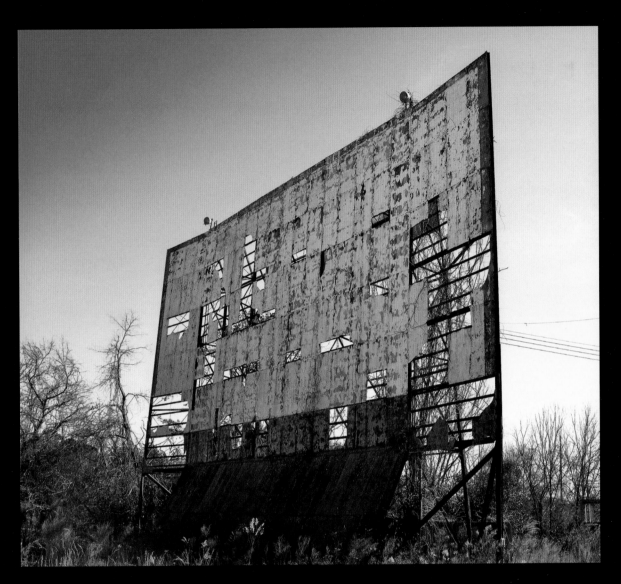

An abandoned drive-in theater in Clanton, Alabama. Wikimedia Commons, photo by Magnolia

(Eventually, autos would have individual speakers, allowing viewers to tune in the soundtrack on their radios.)

Drive-ins steadily spread across the country, hitting a peak of 4,063 businesses operating in 1958. Many catered to families who wanted to save money, and offered playground areas to entertain the kids. Gradually, teens populated the drive-ins, attracted by the privacy they could have inside their vehicles. Many of these outdoor movie venues became known as "passion pits." Today, drive-ins have dwindled to an estimated 330 or so operations across the country. (Check Drive-ins.com for the latest statistics.)

In the spring of 2018, the Circus Drive-In in Wall Township, New Jersey, was demolished after sixty-four years of serving visitors to the Jersey Shore. Photo by Neil Ostrander, used with permission

Poster, Photofest

The 1973 George Lucas film *American Graffiti* centered around Mel's Drive-In, located at 140 South Van Ness in San Francisco. The film captured the nostalgia for the 1950s, when drive-ins were favorite hangouts for teens and young adults. Photofest

Adieu to Drive-In Diners, Too

Just as drive-in theaters have been slowly dying out, so too have drive-in restaurants. At one point, the whole idea of spending more time in your car was more appealing to the American public than it is today. The novelty has worn off. Drive-in restaurants let customers pull into their lots and get served without ever leaving their vehicles. The first drive-in restaurant was Kirby's Pig Stand, which opened in 1921 in Dallas, Texas. The menu included chicken-fried steak sandwiches, deep-fried onion rings, and Texas toast. In time, the concept of drive-ins was replaced by the drive-through, where people could pick up their orders at a window and drive off.

ABOVE: **Barr's Drive-In in San Luis Obispo, California, fed drivers in the 1930s and '40s.** Wikimedia Commons, Boston Public Library Tichnor Brothers Collection

RIGHT: **Andy's Drive-In in Kenosha, Wisconsin, is a classic, and continues to serve juicy burgers to this day. Hot rods and other automobiles from the 1950s and '60s are often parked out front.** Wikimedia Commons, photo

Abandoned Amusement Parks

There's something especially sad and creepy about a deserted amusement park. When the fun ends and the laughter of children fades away, these locations can be downright spooky. While amusement parks like Disneyland and Six Flags are not going anywhere, many local fun parks have been shuttered. Here are a few from around the United States:

After four years of delays, Little Kahuna, a pirate-themed water park was slated to restart construction in 2013, with six water slides, a 920-foot-long lazy river, and a 9,500-square-foot sea-creature-themed kiddie area. The park hasn't yet opened, and it remains abandoned, with all of its attractions in place. Photo by Jonathan Haeber, reprinted with permission

One of the oldest amusement parks in the country, Williams Grove Amusement Park, near Mechanicsburg, Pennsylvania, provided thrills from 1850 until 2005. The wooden Cyclone roller coaster was one of its main attractions. Photo by Jonathan Haeber, reprinted with permission

Neverland, the infamous private amusement park built and owned by singer Michael Jackson was named after the fantasy island in Peter Pan, the story of a boy who never grows up. Since Jackson's death in 2009, the theme-park rides have been removed, and the property rebranded as the ultra-luxurious Sycamore Valley Ranch. Photo by Jonathan Haeber, reprinted with permission

BIBLIOGRAPHY

SECTION 1
LOST KINGDOMS OF EARLY AMERICA: THE MYTHICAL AND THE REAL

Cahokia: North America's First Great City

Allen, David W. "Israelite and Egyptian Connections to Ancient Earthworks near Newark, OH, USA." It's About Time. https://itsabouttimebook.com/newark-ohio-earthworks/

Berg, Emmett. "The Lost City of Cahokia." *Humanities*. September/October 2004. https://www.neh.gov/humanities/2004/septemberoctober/feature/the-lost-city-cahokia

Gidwitz, Tom. "Cities Upon Cities." Archaeology. July/August 2010. https://archive.archaeology.org/1007/abstracts/huasteca.html

Henderson, Harold. "The Rise and the Fall of the Mound People." Chicago Reader. June 29, 2000. https://www.chicagoreader.com/chicago/the-rise-and-fall-of-the-mound-people/Content?oid=902673

Hodges, Glenn. "America's Forgotten City." *National Geographic*. January 2011. https://www.nationalgeographic.com/magazine/2011/01/

Janus, Owen. "Cahokia: North America's First City." LiveScience. January 11, 2018. https://www.livescience.com/22737-cahokia.html

Pauketet, Timothy. *Cahokia: Ancient America's Great City on the Mississippi*. New York: Penguin Library of American Indian History, 2010.

Paulson, Amanda. "The Inca, Maya, and - Cahokian?" The Christian Science Monitor. December 31, 2004. https://www.csmonitor.com/csm/contentmap/articles/2004-12-31

Seppa, Nathan. "Metropolitan Life on the Mississippi." *The Washington Post*. March 12, 1997. http://www.washingtonpost.com/wp-srv/national/daily/march/12/cahokia.htm

Thornton, Richard. "Who Build Cahokia." PeopleofoneFire.com. November 27, 2013. https://peopleofonefire.com/who-built-cahokia.html

Quivira and the Seven Golden Cities of Cibola

The Fountain of Youth

"Coronado's Journey Through New Mexico, Texas, Oklahoma, and Kansas." Planetary Science Institute. http://www.psi.edu/about/staff/hartmann/coronado/coronadosjourney2.html

"The Coronado Expedition." The Arizona Experience. http://arizonaexperience.org/remember/coronado-expedition

"Francisco Vazquez de Coronado." History.com https://www.history.com/topics/exploration/francisco-vazquez-de-coronado

"Juan Ponce de León." Encyclopedia Brittanica. https://www.britannica.com/biography/Juan-Ponce-de-Leon

"Radioactive Fountain of Youth." RoadsideAmerica.com. June 17, 2018. https://www.roadsideamerica.com/story/47431

Anderson, Sam. "Searching for the Fountain of Youth." *The New York Times*. October 24, 2014. https://www.nytimes.com/2014/10/26/magazine/my-search-for-the-fountain-of-youth.html

Drye, Willie. "Seven Cities of Cibola." National Geographic. https://www.nationalgeographic.com/archaeology-and-history/archaeology/seven-cities-of-cibola/

Greenspan, Jesse. "The Myth of Ponce de Leon and the Fountain of Youth." History.com. April 2, 2013. https://www.history.com/news/the-myth-of-ponce-de-leon-and-the-fountain-of-youth

O'Neill, Natalie. "Teen accidentally helps discover lost 16th-century civilization in Kansas." *The New York Post*. April 18, 2017. https://nypost.com/2017/04/18/teen-accidentally-helps-discover-lost-16th-century-civilization/

Sharp, Jay W. "Coronado Expedition: From Cibola to Quivira."

DesertUSA. https://www.desertusa.com/desert-trails/coronado-expedition-quivira.html

Weiser, Kathy. "The Kingdom of Quivira, Kansas." Legends of America. July 2018 http://www.legendsofkansas.com/quivira.html

Norumbega: The Viking City of New England

"Humphrey Gilbert." Encyclopedia Britannica. https://www.britannica.com/biography/Humphrey-Gilbert

Browne, Patrick. "Norumbega: Did the Vikings Beat the Pilgrims to Plymouth?" Historical Digression. July 24, 2014. https://historicaldigression.com/tag/norumbega/

Holloway, April. "Do Spirit Pond Inscriptions show that the Holy Grail was taken to North America?" Ancient Origins: Reconstructing the Story of Humanity's Past. March 15, 2014. http://www.ancient-origins.net/news-general/do-spirit-pond-inscriptions-show-holy-grail-was-taken-north-america-001449

Lentini, Grace. "The Great Narragansett Rune Stone Debate." SORhodeIsland.com. November 24, 2015. http://sorhodeisland.com/stories/The-Great-Rune-Stoen-Debate,17149

MacKay, Art. "MYSTERIES: Can you find the lost city of Norumbega . . . Is it somewhere in New England?" Medium. April 5, 2017. https://medium.com/@artmackay/mysteries-can-you-find-the-lost-city-of-norumbega-is-it-somewhere-in-new-england-67c79fb01ecb

Ogbur, Charlton. "The Longest Walk: David Ingram's Amazing Journey." American Heritage. April/May 1979. https://www.americanheritage.com/content/longest-walk-david-ingram%E2%80%99s-amazing-journey

Tooohey, John. "The Long, Forgotten Walk of David Ingram." *The Public Domain Review.* https://publicdomainreview.org/2017/06/28/the-long-forgotten-walk-of-david-ingram/

Whitney. Emmie Bailey. "The Lost City of Norumbega." American History and Genealogy Project. August 2011. http://www.ahgp.org/maine/lost-city-of-norumbega.html

Woodruff, Andy. "Norumbega, New England's lost city of riches and Vikings." Andy Woodruff Cartographer/Blog. May 24, 2010. http://andywoodruff.com/blog/norumbega-new-englands-lost-city-of-riches-and-vikings/

SECTION 2
ABANDONED AMERICA: COMMUNITIES THAT FLOURISHED AND FADED

The Vast Anasazi Civilization of the Southwest

"The Anasazi." Northern Arizona University. http://dana.ucc.nau.edu

"The Lost City." National Park Service. https://www.nps.gov/lake/learn/the-lost-city.htm

"Native American Heritage." Scholastic. http://teacher.scholastic.com/researchtools/research-starters/native_am/

Blinman, Eric. "Anasazi Potter: Evolution of a Technology." Penn Museum/*Expedition Magazine* 35.1 (1993). https://www.penn.museum/sites/expedition/anasazi-pottery/

La Porte, John. "Anasazi People Did Not Mysteriously Disappear." The Fort Morgan Times. May 14, 2010. http://www.fortmorgantimes.com/ci_15083924

Wiener, James. "The Mysterious Ancient Puebloan Peoples (Anasazi)." Ancient History Et Cetera. September 19, 2012. http://etc.ancient.eu/interviews/interview-the-ancient-anasazi-in-focus/

Roanoke: America's First Colony

Evans, Phillip. "Amadas and Barlowe Expedition." NCPedia. 2006. https://www.ncpedia.org/amadas-and-barlowe-expedition

Horn, James. *A Kingdom Strange: The Brief and Tragic History of the Lost Colony of Roanoke.* New York: Basic Books/Hachette, 2011.

Horn, James. "Roanoke's Lost Colony Found?" American Heritage. Spring 2010. https://www.americanheritage.com/content/roanokes-lost-colony-found

Jones, Justin. "What Happened

to Roanoke's 'Lost Colonists'?" Daily Beast. August 12, 2015. https://www.thedailybeast.com/what-happened-to-roanokes-lost-colonists

Tondu, Gerard. "1584 – Amadas & Barlowe." U.S. Timeline. August 20, 2013. https://sites.google.com/site/atimelineofamerica/1584---amadas-barlowe

Wolfe, Brendan. "The Roanoke Colonies." Encyclopedia Virginia. June 13, 2014. https://www.encyclopediavirginia.org/Roanoke_Colonies_The

Bodie: One of America's Most Famous Ghost Towns

"Bodie: A Ghostly Ghost Town." Legends of America." https://www.legendsofamerica.com/ca-bodie/3/

"Madame Moustache- The poignant story of the fearless woman, known to be the best blackjack dealer in the Old West." *Vintage News.* February 8, 2016. https://www.thevintagenews.com/2016/02/08/madame-moustache/

"Mono Mills." The Historical Marker Database. https://www.hmdb.org/marker.asp?marker=50144

Al-Othman, Hannah. "A real ghost town: Inside the abandoned Wild West community of Bodie which sprung up with the Gold Rush and died again just as quickly." DailyMail.com. November 4, 2016. http://www.dailymail.co.uk/news/article-3905064/A-real-ghost-town-Inside-abandoned-Wild-West-community-Bodie-sprung-Gold-Rush-died-just-quickly.html

Hausladen, Gary. *Western Places, American Myths: How We Think about the West.* Reno: University of Nevada Press, 2006.

Hoeller, Sophie Claire. "12 of the Most Eerily Abandoned Towns in America." Thrillist. https://www.thrillist.com/travel/nation/abandoned-towns-in-america-10-of-most-eerily-ghost-towns

McNair, Josh. "Bodie State Historic Park: California's Best Ghost Town." California Through My Lens. https://californiathroughmylens.com/bodie-ghost-town

Nickell, Joe. "The Curse of Bodie: Legacy of Ghost-Town Ghosts?" CSI. November/December 2003. https://www.csicop.org/si/show/curse_of_bodie_legacy_of_ghost-town_ghosts

Piatt, Michael. "Bodie, California: Myth Blasters and Articles About Bodie's History & Technology." Bodie, California: History and Research. http://www.bodiehistory.com/essays.htm
Piatt, Michael. "The Death of Madame Mustache: Bodie's Most Celebrated Inhabitant." Bodie, California: History and Research. March 2009. http://www.bodiehistory.com/madame.htm

Rhodes, Catie. "The Curse That Follows You." CatieRhodes.com. https://catierhodes.

com/2012/04/the-curse-that-follows-you/

Strochlic, Nina. "America's Tiniest Town Is Sold and Renamed Phindeli Town Buford, Wyoming." Daily Beast. October 17, 2013. https://www.thedailybeast.com/americas-tiniest-town-is-sold-and-renamed-phindeli-town-buford-wyoming

North Brother Island: New York Center for Quarantining the Sick

"North Brother Island." Atlas Obscura. https://www.atlasobscura.com/places/north-brother-island

"Typhoid Mary within a Contact Network." Cornell University. November 29, 2015. https://blogs.cornell.edu/info2040/2015/11/29/typhoid-mary-within-a-contact-network/

Cunha, Darlena. "A Condensed History of Quarantine's Success and Failure." Atlas Obscura. October 24, 2014. https://www.atlasobscura.com/articles/a-condensed-history-of-quarantines-success-and-failure

Foderaro, Lisa. "On an Island Under Vines, New York City Officials See a Future Park." *The New York Times.* October 15, 2014. https://www.nytimes.com/2014/10/16/nyregion/on-an-island-under-vines-officials-see-a-future-park.html

Klein, Christopher. "10 Things You May Not Know About "Typhoid Mary"." History.com March 27, 2015. https://www.history.

com/news/10-things-you-may-not-know-about-typhoid-mary

Klibanoff, Eleanor. "Awful Moments In Quarantine History: Remember Typhoid Mary?" NPR. October 30, 2014. https://www.npr.org/sections/goatsandsoda/2014/10/30/360120406/awful-moments-in-quarantine-history-remember-typhoid-mary

Mosher, Dave. "New York City owns a creepy island that almost no one is allowed to visit — here's what it's like." Business Insider. October 8, 2017. http://www.businessinsider.com/north-brother-island-photo-tour-2017-9

Robin, Josh. "The Secret of North Brother Island: The Abandoned New York City Island Where Typhoid Mary Was Held Captive." *Daily Beast.* December 29, 2017. https://www.thedailybeast.com/the-secret-of-north-brother-island-the-abandoned-new-york-city-island-where-typhoid-mary-was-held-captive

Smith, Kiona N. "Who Was Typhoid Mary?" Forbes. September 22, 2017. https://www.forbes.com/sites/kionasmith/2017/09/22/who-was-typhoid-mary/2/#16fb3bb6cad7

Tyson, Peter. "A Short History of Quarantine." PBS/NOVA. October 12, 2014. http://www.pbs.org/wgbh/nova/body/short-history-of-quarantine.html

Williams, Timothy. "City Claims Final Private Island in East River." *The New York Times.* November 20, 2007. https://www.

nytimes.com/2007/11/20/nyregion/20brother.html

SECTION 3
MOTHER NATURE STRIKES: TOWNS LOST TO NATURAL DISASTER

Ruddock, Napton, and Frenier, Louisiana: Wiped Out by Hurricane

Isle Derniere, Louisiana
Indianola, Texas
Hog Island, New York

"6 Hurricane Ghost Towns." The Weather Channel. August 9, 2013. https://weather.com/tv/shows/hurricane-week/news/hurricane-ghost-towns-photos-20130807#/5

"Frenier Cemetery." https://www.findagrave.com/cemetery/2562811/frenier-cemetery

"Heart of Louisiana: Hurricane of 1915." Fox8. October 27, 2016. http://www.fox8live.com/story/33502626/heart-of-louisiana-hurricane-of-1915

"Hurricane submerges Louisiana Resort." History.com. https://www.history.com/this-day-in-history/hurricane-submerges-louisiana-resort

"Indianola, Texas." Indianolatx.com. http://www.indianolatx.com/history.html

"Old Cahawba Archaeological Site." ExploreSouthernHistory.com. http://www.exploresouthernhistory.com/oldcahawba.html

"Ruddock." Ghosttowns.com. http://www.ghosttowns.com/states/la/ruddock.html

"Ruddock washed away 100 years ago." L'Observateur. September 30, 2015. https://www.lobservateur.com/2015/09/30/ruddock-washed-away-100-years-ago/

"What really happened to Hog Island?" New York Historical Society Museum & Liubrary. August 31, 2011. http://blog.nyhistory.org/what-really-happened-to-hog-island/

Bellot, Alfred. *History of the Rockaways.* New York: Bellot's History of the Rockaways, Inc., 1917

Campanella, Richard. *Cityscapes of New Orleans.* Baton Rouge, Louisiana: LSU Press, 2017

Elrick, Wil. "Old Cahawba–Alabama's capital Ghost Town." Our Valley Events. June 10, 2016. https://ourvalleyevents.com/old-cahawba-alabamas-capital-ghost-town/

Frishberg, Hannah. "8 Long Lost Islands That Used To Be Part of New York City." Curbed: New York. December 3, 2014. https://ny.curbed.com/2014/12/3/10016116/8-long-lost-islands-that-used-to-be-part-of-new-york-city

Herndon, Ernest. "Old cemetery, railroad hint at Ruddock's history. Little remains of once-thriving settlement." Enterprise-Journal. July 15, 2017. http://www.enterprise-journal.com/sports/outdoors/article_24df-

8b1a-699a-11e7-901a-d353efde-0ca3.html

McCloskey, James. "Hog Island." Underwater New York. September 6, 2013. http://underwater-newyork.com/archive/2013/9/6/hog-island-by-james-mcclosky

Mcilvain, Myra. "The Rise and Fall of Indianola." Myrahmcilvain.com. https://myrahmcilvain.com/2017/07/21/the-rise-and-fall-of-indianola/

Norwood, Wayne. *The Day Time Stood Still: The Hurricane of 1915.* North Charleston, South Carolina: CreateSpace Independent Publishing Platform, 2015.

Onishi, Norimitsu. "Queens Spit Tried to Be a Resort but Sank in a Hurricane." *The New York Times.* March 18, 1997. https://www.nytimes.com/1997/03/18/nyregion/queens-spit-tried-to-be-a-resort-but-sank-in-a-hurricane.html

Thompson, Dave. *Bayou Underground: Tracing the Mythical Roots of American Popular Music.* Toronot, Canada: ECW Press, 2010.

Trickey, Erick. "A Hurricane Destroyed This Louisiana Resort Town, Never to Be Inhabited Again: The destruction of Isle Derniere resonates as history's warning for our era of rising seas." Smithsonian.com. January 4, 2017. https://www.smithsonianmag.com/history/hurricane-destroyed-louisiana-resort-town-never-be-inhabited-again-180961645/

Vanport, Oregon: Flood Washes Away Oregon's Second Largest City

Old Cahawba, Alabama Drowned Towns of the Catskills.

"The Submerged Towns of America." Uncharted101. July 31, 2015. https://www.uncharted101.com/the-submerged-towns-of-america/

"Vanport Residences, 1947." The Oregon History Project. https://oregonhistoryproject.org/articles/historical-records/vanport-residences-1947/#.WygSA-BlJmL1

Geiling, Natasha. "How Oregon's Second Largest City Vanished in a Day." Smithsonian.com February 18, 2015. https://www.smithsonianmag.com/history/vanport-oregon-how-countrys-largest-housing-project-vanished-day-180954040/

Kemble, William J. Ashokan. "Reservoir construction completed 100 years ago today." *Daily Freeman.* June 23, 2014. http://www.dailyfreeman.com/general-news/20140623/ashokan-reservoir-construction-completed-100-years-ago-today

Loftin, A.J. "The Ashokan Reservoir: The creation of the Ashokan Reservoir changed the Catskills forever." *Hudson Valley.* July 17, 2008. http://www.hvmag.com/Hudson-Valley-Magazine/August-2008/History-The-Ashokan-Reservoir/

Oliver, Gordon. "Kaiser Shipyards." The Oregon Encyclopedia. https://oregonencyclopedia.org/articles/kaiser_shipyards/#.WygQ0xlJmL1

Rubenstein, Sura "1998: Flood of Change." Oregonlive.com. December 11, 2014. https://www.oregonlive.com/history/2014/12/1998_story_flood_of_change.html

Salazar, Cristian. "How New York City gets its water, from reservoir to tap: NYCurious." *amNewYork.* April 18, 2018. https://www.amny.com/lifestyle/how-nyc-gets-its-water-1.9205765

Shenoy, Rupa. "In the shadow of a racist past, Portland still struggles to be welcoming to all its residents." PRI. October 26, 2017. https://www.pri.org/stories/2017-10-26/shadow-racist-past-portland-still-struggles-be-welcoming-all-its-residents

Young, Michelle. "Lots of NYC's Drinking Water Comes from Drowned Towns in the Catskills." Untapped Cities. June 22, 2015. https://untappedcities.com/2015/06/22/some-of-nycs-drinking-water-comes-from-drowned-towns-in-the-catskills/

SECTION 4
MAN-MADE DISASTERS: TOWNS TOO POLUTED TO SURVIVE

Love Canal: America's First Toxic Ghost Town

Beck, Eckhardt. "The Love Canal Tragedy." *EPA Journal.* January 1979. https://archive.epa.gov/epa/aboutepa/love-canal-tragedy.html

Brook, Marissa. "The Tragedy of Love Canal." Damn Interesting. October 18, 2006. https://www.

damninteresting.com/the-trage-dy-of-the-love-canal/

Brown, Michael. "A Toxic Ghost Town." *The Atlantic*. July 1989. https://www.theatlantic.com/magazine/archive/1989/07/a-tox-ic-ghost-town/303360/

Gibbs, Lois Marie. "History: Love Canal: the Start of a Movement." Boston University. https://www.bu.edu/lovecanal/canal/

Johnson, George. "How Many People Were Killed by Love Canal?" Slate.com. august 28, 2013. http://www.slate.com/articles/health_and_science/medical_examiner/2013/08/love_canal_killed_how_many_people_cancer_risk_from_environmental_pollution.html

Kleiman, Jordan. "Love Canal: A Brief History." SUNY Geneseo. https://www.geneseo.edu/history/love_canal_history.

Revkin, Andrew. "Love Canal and Its Mixed Legacy." The New York Times. November 25, 2013. https://www.nytimes.com/2013/11/25/booming/love-canal-and-its-mixed-legacy.html

Thompson, Carolyn. "Lawsuits: Love Canal still oozes 35 years later." *USAToday*. November 2, 2013. https://www.usatoday.com/story/money/business/2013/11/02/suits-claim-love-canal-still-oozing-35-years-later/3384259/

Verhovek, Sam Howe. "After 10 Years, the Trauma of Love Canal Continues." *The New York Times*.

August 5, 1988. https://www.nytimes.com/1988/08/05/nyregion/after-10-years-the-trauma-of-love-canal-continues.html

Centralia: A Town Consumed by a Coal Mining Fire

"Route 61 (Centralia)." Dangerousroads: The world's most spectacular roads. http://www.dangerousroads.org/north-america/usa/3843-abandoned-route-61-centralia.html

"Venting the Centralia, Pennsylvania Mine Fire." Centralia.org. October 1, 2014. http://www.centraliapa.org/venting-centra-lia-pennsylvania-mine-fire/

Andrews, Evan. "9 Things You May Not Know About the Oregon Trail." History.com November 13, 2015. https://www.history.com/news/9-things-you-may-not-know-about-the-oregon-trail

Beck, Graham T. "When the Town Stops Burning." Themorningnews.org. https://themorningnews.org/article/when-the-town-stops-burning

Dekok, David. *Fire Underground: The Ongoing Tragedy of the Centralia Mine Fire*. Guilford, Connecticut: Globe Pequot Press, 2009. Eveleth, Rose. "Why People Won't Leave the Town that Has Been On Fire for Fifty Years." *Smithsonian Magazine*. August 10, 2012. https://www.smithsonianmag.com/smart-news/why-people-wont-leave-the-town-that-has-been-on-fire-for-fifty-years-20041449/

Hurt, Harry. "Chronicling and Underground Inferno." *The New York Times*. May 20, 2007. http://www.nytimes.com/2007/05/20/business/yourmoney/20shelf.html

Kalson, Sally. "Slow Burn in Centralia, PA." *The New York Times*. November 22, 1981.

Morton, Mary Caperton. "Hot as Hell: Firefighting foam heats up coal fire debate in Centralia, Pa." *Earth Magazine*. May 5, 2010. https://www.earthmagazine.org/article/hot-hell-firefighting-foam-heats-coal-fire-debate-centralia-pa

Palus, Shannon. "Extreme Science: Town on Fire." *Popular Science*. July 15, 2015. https://www.popsci.com/extreme-science-centralia

Quigley, Joan. *The Day the Earth Caved In: An American Mining Tragedy*. New York: Random House, 2009.

Quigley, Joan. "Pictures: Centralia Mine Fire, at 50, Still Burns With Meaning." *National Geographic*. January 10, 2013. https://news.nationalgeographic.com/news/energy/2013/01/pictures/130108-centralia-mine-fire/

Walter, Greg. "A Town with a Hot Problem Decides Not to Move Mountains but to Move Itself." *People* magazine. June 22, 1981. https://people.com/archive/a-town-with-a-hot-problem-decides-not-to-move-mountains-but-to-move-itself-vol-15-no-24/

Wood, Lawrence. *One Hundred Tons of Ice: And Other Gospel Stories*. Louisville, Kentucky: Westminster John Knox Press, 2004.

Gilman: A Deadly Water Supply Drives a Town to Extinction

"Gilman, Colorado." Substreet. https://substreet.org/gilman-colorado/

"Gilman, Colorado Ghost Town." Uncover Colorado. https://www.uncovercolorado.com/ghost-towns/gilman/

"The History and Future of the Eagle Mine." Vail Symposium. August 1, 2017. http://vailsymposium.org/th_event/eagle-mine/

"Searching for Ghosts in Gilman, Colorado." Ski Travel. https://skjtravel.net/index.php/46-uncategorised/305-ghost-town-gilman-colorado

Boyd, Pam. "Cleanup at the long-closed Eagle Mine near Minturn will probably never end." Vail Daily. August 4, 2017. https://www.vaildaily.com/news/cleanup-at-the-long-closed-eagle-mine-near-minturn-will-probably-never-end/

Gibson, Mark. "Viewpoint: Leveraging EPA's Orange River to Abate the Threat of Abandoned Mines." Your Colorado Water Blog. August 20, 2015. https://blog.yourwatercolorado.org/2015/08/20/viewpoint-leveraging-epas-orange-river-to-abate-the-threat-of-abandoned-mines/

Schnell, Caramie. "A forgotten town." *Vail Daily*. June 14, 2010. https://www.vaildaily.com/news/a-forgotten-town/

Thompson, Cliff. "Eagle River cleanup going public." *Vail Daily*. November 19, 2002. https://www.vaildaily.com/news/eagle-river-cleanup-going-public/

Tobias, Jimmy. "Meet the Toxic Time Bombs Hidden in the High Country." *Outside*. August 18, 2015. https://www.outsideonline.com/2009316/meet-toxic-time-bombs-hidden-high-country

SECTION 5
FADING ICONIC AMERICA: LOST HIGHWAYS, DRIVE-INS, AND DINERS

Route 66
The Oregon Trail
The Lincoln Highway

Drive-ins
Amusement Parks

"Historic Route 66 Travel Guide." Route66Guide. https://www.route66guide.com/

"The History of Route 66." National Historic Route 66 Federation. https://www.national66.org/history-of-route-66/
"Oregon Trail." History.com. https://www.history.com/news/9-things-you-may-not-know-about-the-oregon-trail
"Route 66." Road Trip USA. https://roadtripusa.com/route-66/

Bliss, Laura. "The End of America's Love Affair with Route 66." Citylab. December 2, 2014. https://www.citylab.com/transportation/2014/12/the-end-of-americas-love-affair-with-route-66/383335/

Crapanzano, Christina. "Route 66." *Time*. June 28, 2010. http://content.time.com/time/nation/article/0,8599,2000095,00.html

Kapotes, Emma. "8 Beautifully Haunting Pictures of Route 66 Ghost Towns." *Reader's Digest*. https://www.rd.com/advice/travel/ghost-town-pictures/

Lamb, David. "The Mystique of Route 66." *Smithsonian Magazine*. February 2012. https://www.smithsonianmag.com/travel/the-mystique-of-route-66-19433629/

Lin, James. "A Brief History: Origins, 1912-13." Lincoln Highway. http://lincolnhighway.jameslin.name/history/part1.html

Reid, Robert. "Road Trip: Route 66." *National Geographic*. April 4, 2014. https://www.nationalgeographic.com/travel/road-trips/route-66/

Weiser, Kathy. "Glenrio—One More Route 66 Casualty." Legends of America. March 2017. https://www.legendsofamerica.com/tx-glenrio/

Ronca, Debra. "What's so special about Route 66?" HowStuffWorks. https://adventure.howstuffworks.com/route-66.htm

Weingroff, Richard. "The Lincoln Highway." Federal Highway Administration. https://www.fhwa.dot.gov/infrastructure/lincoln.cfm

ABOUT THE AUTHOR

DON RAUF has written more than 40 nonficton books, covering historical topics such as the French and Indian War, the Mexican American War, and Washington's Farewell Address. His book Schwinn: The Best Present Ever (Lyons Press) took a detailed look at the history of Schwinn and the influence the brand has had on American culture. Don lives in Seattle. ∎

ACKNOWLEDGEMENTS

Thanks to Rick Rinehart for bringing this project my way and guiding it to completion. At Globe Pequot, I am also grateful to Alex Bordelon and Stephanie Scott for seeing it through the final stages of production. Also, thank you to the photographers who supplied original images, especially Jonathan Haeber and Scott MX Turner.

Finally my thanks to A & E Networks and the History Channel for vetting the book and providing their enthusiastic support, especially Paula Allen, Jill Tully, Naz Atlan, and Kim Gilmore.∎

Six Flags New Orleans was abandoned following Hurricane Katrina in 2005. Though off-lim-
its to the public, the park was nevertheless
used in scenes from *Jurassic World* (2015) and
Dawn of the Planet of the Apes (2014). Wikimedia Commons